The Aging Musculoskeletal System

 **The Collamore Press**
D. C. Heath and Company
Lexington, Massachusetts
Toronto

The Aging Musculoskeletal System

Physiological and Pathological Problems

Edited by

CARL L. NELSON, M.D.
University of Arkansas for Medical Sciences
Little Rock, Arkansas

ANTHONY P. DWYER, M.D.
Ochsner Clinic
New Orleans, Louisiana

Every effort has been made to ensure that drug dosage schedules and indications are correct at time of publication. Since ongoing medical research can change standards of usage, and also because of human and typographical error, it is recommended that readers check the *PDR* or package insert before prescription or administration of the drugs mentioned in this book.

Copyright © 1984 by D.C. Heath and Company

All rights reserved. No part of this publication may be reproduced or transmitted in any form or by any means, electronic or mechanical, including photocopy, recording, or any information storage or retrieval system, without permission in writing from the publisher.

Published simultaneously in Canada

Printed in the United States of America

International Standard Book Number: 0-669-05653-7

Library of Congress Catalog Card Number: 84-70464

Library of Congress Cataloging in Publication Data
Main entry under title:

The Aging musculoskeletal system.

Papers from a conference, Sept. 28–30, 1981, organized by the Dept. of Orthopaedic Surgery, University of Arkansas.
Includes index.
1. Musculoskeletal system—Diseases—Age factors—Congresses. 2. Musculoskeletal system—Aging—Congresses. 3. Bones—Diseases—Age factors—Congresses. I. Nelson, Carl L. II. Dwyer, Anthony P. III. University of Arkansas for Medical Sciences. Dept. of Orthopaedic Surgery. [DNLM: 1. Aging. 2. Musculoskeletal System. WE 100 A267]
RC925.5.A45 1984   618.97'67   84-70464
ISBN 0-669-05653-7

Contents

◀

Preface *Anthony P. Dwyer, M.D., and Carl L. Nelson, M.D.*	vii
Introduction *Eugene J. Towbin, M.D., Ph.D.*	ix
Contributors	xi
1 Overview: The Demographics of Aging *Robert B. Hudson, Ph.D.*	1
2 Psychosocial Factors in Aging *Owen W. Beard, M.D.*	9
3 The Biology and Physiology of Aging *Edward J. Masoro, Ph.D.*	17
4 Pathoanatomy of Aging: An Overview *Aubrey J. Hough, Jr., M.D.*	33
5 The Biology of Aging Human Collagen *LeRoy Klein, M.D., Ph.D., and Jess C. Rajan, Ph.D.*	37
6 The Biology of Aging Muscle *Peter Jokl, M.D.*	49
7 The Aging of Articular Cartilage *O. Donald Chrisman, M.D.*	59
8 The Biology of Aging Bone *Howard Duncan, M.D., and A. Michael Parfitt, M.B.*	65
9 Osteoporosis *Jenifer Jowsey, Ph.D.*	75
10 The Kinematics of Aging *R. Donald Hagan, Ph.D.*	91
11 The Aging Lumbar Spine *Dan M. Spengler, M.D.*	103
12 Current Concepts of the Pathogenesis of Osteoarthritis *Leon Sokoloff, M.D.*	113
13 Early Aging Nutritional Changes at the Base of the Articular Cartilage *Darrel W. Haynes, M.S., Ph.D.*	121

14 Biomechanics of Joint Deterioration and Osteoarthritis   127
    Eric L. Radin, M.D., and R. Bruce Martin, Ph.D.

            15 Total Hip Replacement in the Elderly   135
                Carl L. Nelson, M.D.

                    16 Trauma in the Elderly   141
    Gordon A. Hunter, M.B., and E.T.R. James, M.B.

                17 Hip Fractures in the Elderly   147
    Anthony P. Dwyer, M.D., and Carl L. Nelson, M.D.

              18 Rehabilitation Medicine in Aging   157
                William J. Erdman II, M.D.

    19 The Aging Population: Prosthetic and Orthotic
                                      Considerations   163
                    John H. Bowker, M.D.

                             Index   171

Preface

As the "graying of America" continues, physicians are faced with increasing demands to maintain a high quality of life for the growing number of patients who both live longer and remain physically active, even athletic, at advanced ages. Improving health care for the elderly will depend on progress in understanding and treating the difficult musculoskeletal problems they encounter: loss of joint function, joint stiffness, pain from degenerative joint disease, and the fragility of osteoporotic bone. As our society's demographics shift, in conjunction with our pursuit of sport and exercise, these disorders will become an ever more important aspect of health care.

The other body systems also age and deteriorate. As the aging human body grows more susceptible to musculoskeletal problems, it becomes less able to cope with the extra stresses that result from compensating for loss of joint function or from the surgical trauma and immobilization involved in treating bone and joint disorders. Therefore, this volume also covers nonorthopedic clinical problems, particularly in rehabilitation, as well as the psychosocial and demographic factors of which physicians must be aware in treating aging patients.

In covering all of these topics—the basic and clinical science, the psychosocial and rehabilitative factors—we hope this book will serve as a step toward greater understanding of and improved care for the aging musculoskeletal system.

*Teach us to live that we may dread
Unnecessary time in bed.
Get people up and we may save
Our patients from an early grave.*

Richard Asher

Introduction

In this day of great enthusiasm for the role of exercise in the maintenance of health, there has been considerable interest in the use of exercise by that rapidly growing segment of the population: the aged. Evidence of this interest is apparent in the many elderly seen jogging, on tennis courts, or doing calisthenics all over this nation. This enthusiasm is so widespread that amateur Masters Running events are very popular and professional Masters Tennis competitions that have a minimal entry age of 40 are financially successful. With the frail elderly, physical reconditioning has been used as a vehicle for resocialization, as therapy for depression, and for other purposes that may be fitted under the broad heading of improving the quality of life. On a more biochemical and physiologic level of geriatrics, much has been written about the effects of exercise on cardiopulmonary functions, vascular atheromatous disease, and obesity with or without maturity-onset diabetes.

The use of exercise for preventive, restorative, or maintenance purposes depends on a reasonably functional musculoskeletal system. Those geriatric patients with seriously compromised musculoskeletal systems frequently cannot shop, travel for health care, or otherwise maintain an independent life-style and are forced to seek institutional care. In fact, problems with locomotion, mentation, and incontinence account for a large majority of nursing home placements. Recourse to institutional care is not only a frightfully expensive solution, but too frequently is attended by depression, isolation, and unhappy disintegration.

The Veterans Administration has shown imaginative and dynamic leadership in developing special programs of clinical service for the elderly veteran and special centers to facilitate teaching and research in geriatrics. There are now nine such Geriatric Research, Education and Clinical Centers (GRECC) distributed around the country in VA hospitals closely affiliated with colleges of medicine. The Little Rock Veterans Administration Medical Center has been the site of an active GRECC since 1974. Intellectually stimulated by this focus on geriatric problems, the Department of Orthopaedic Surgery of the University of Arkansas for Medical Sciences organized a two-day conference, September 28–30, 1981, to review the basic science and clinical knowledge regarding the aging musculoskeletal system. Such a multidisciplinary conference represented our best hopes to allay and treat the ravages wrought by time and disease on our abilities for locomotion. It also represented a fine example of the mutually beneficial partnership between a college of medicine and a Veterans Administration hospital.

Eugene J. Towbin, M.D., Ph.D.

Contributing Authors

Owen W. Beard, M.D.
Chief, Division of Geriatrics
Little Rock Veterans Administration Medical Center
Professor of Medicine
University of Arkansas School of Medicine
Little Rock, Arkansas

John H. Bowker, M.D.
Director, Rehabilitation Center
Jackson Memorial Hospital
Professor, Department of Orthopaedics and Rehabilitation
University of Miami School of Medicine
Miami, Florida

O. Donald Chrisman, M.D.
Clinical Professor, Surgery (Orthopaedics)
Yale University School of Medicine
New Haven, Connecticut

Howard Duncan, M.D., F.R.C.P., F.A.C.P.
Head, Division of Rheumatology
Henry Ford Hospital
Detroit, Michigan
Clinical Professor of Medicine
University of Michigan

Anthony P. Dwyer, M.D.
Chairman, Orthopaedic Department
Ochsner Clinic
New Orleans, Louisiana

William J. Erdman II, M.D.
Professor and Chairman, Department of Physical Medicine
  and Rehabilitation
University of Pennsylvania
Philadelphia, Pennsylvania

R. Donald Hagan, Ph.D.
Director of Research
Institute for Human Fitness
Texas College of Osteopathic Medicine
Fort Worth, Texas

Darrel W. Haynes, M.S., Ph.D.
Director, Orthopaedic Research
Department of Orthopaedic Surgery
University of Arkansas for Medical Sciences
Little Rock, Arkansas

Aubrey J. Hough, Jr., M.D.
Pathologist in Chief
University Hospital
Professor and Chairman, Department of Pathology
University of Arkansas for Medical Sciences
Little Rock, Arkansas

Robert B. Hudson, Ph.D.
Associate Professor of Social Policy
Fordham University Graduate School of Social Service
New York, New York

Gordon A. Hunter, M.B., F.R.C.S., F.R.C.S.(C)
Orthopaedic Surgeon
Sunnybrook Medical Centre
Associate Professor of Surgery
University of Toronto
Toronto, Ontario

E.T.R. James, M.B., F.R.C.S.
Clinical Fellow in Orthopaedic Surgery
Sunnybrook Medical Centre
Toronto, Ontario

Peter Jokl, M.D.
Director, Athletic Medicine Section/Orthopaedic Surgery
Associate Clinical Professor, Surgery (Orthopaedic)
Yale University School of Medicine
New Haven, Connecticut

Jenifer Jowsey, Ph.D.
Adjunct Professor of Physiology
University of California, Davis
Davis, California

Contributing Authors

LeRoy Klein, M.D., Ph.D.
Assistant Orthopaedist, Department of Orthopaedics
University Hospitals of Cleveland
Professor, Departments of Orthopaedics and
  Macromolecular Science
Case Western Reserve University School of Medicine
Cleveland, Ohio

R. Bruce Martin, Ph.D.
Professor, Departments of Orthopedic Surgery and
  Mechanical and Aerospace Engineering
West Virginia University Medical Center
Morgantown, West Virginia

Edward J. Masoro, Ph.D.
Professor and Chairman, Department of Physiology
University of Texas Health Sciences Center
San Antonio, Texas

Carl L. Nelson, M.D.
Professor, Chairman, and Head
Section of Reconstructive Surgery
Department of Orthopaedic Surgery
University of Arkansas for Medical Sciences
Little Rock, Arkansas

A. Michael Parfitt, M.B., F.R.C.P., F.A.C.P., F.R.A.C.P.
Director of Bone and Mineral Research
Henry Ford Hospital
Detroit, Michigan
Clinical Professor of Medicine
University of Michigan

Eric L. Radin, M.D.
Professor and Chairman, Department of Orthopaedic
  Surgery
West Virginia University Medical Center
Morgantown, West Virginia

Jess C. Rajan, Ph.D.
Assistant Professor, Department of Orthopaedics
Case Western Reserve University School of Medicine
Cleveland, Ohio

Leon Sokoloff, M.D.
Attending Pathologist
University Hospital
Professor of Pathology
State University of New York at Stony Brook
Stony Brook, New York

Dan M. Spengler, M.D.
Professor and Chairman
Department of Orthopaedics and Rehabilitation
Vanderbilt University Medical Center
Nashville, Tennessee

Eugene J. Towbin, M.D., Ph.D.
Chief of Staff and Director of Geriatric Research,
    Education and Clinical Center (GRECC)
Veterans Administration Medical Center
Little Rock, Arkansas
Associate Dean, College of Medicine
University of Arkansas

The Aging Musculoskeletal System

Overview: The Demographics of Aging

Robert B. Hudson, Ph.D.

A number of trends currently underway—demographic, economic, social, and political—are going to alter the place of older persons in American society and the manner in which relations between older persons and society are both structured and played out. It is not possible to predict with certainty future birth rates, economic conditions, or political perceptions, but a combination of extrapolation and informed speculation yields several significant conclusions about the likely standing of older persons in the years ahead.

THE DEMOGRAPHIC PICTURE

Several population developments point to the growing import of older persons in American society. First, the proportion of older persons in the population has already increased notably during this century. In 1900, there were 3.1 million persons 65 years and over, constituting 4.0 percent of the population; in 1930, the 6.6 million Americans 65 years and over composed 5.4 percent of the population; by 1970, the number of older Americans had increased to 20 million, at which time they made up 9.8 percent of the overall population; and, finally, in 1980, there were 25.5 million elderly, equaling 11.3 percent of the total population.

After roughly a 20-year period between now and early in the next century, these numerical and proportional increases will jump again noticeably. Thus, the year 2000 will see a moderate increase in numbers (to about 35 million) and percentage (about 13.1%) of older persons. By 2030, however, the situation will be far different. The post–World War II generation will have aged, and at least the cohorts immediately following them will be significantly smaller in numbers. The over-65 population will be on the order of 65 million persons, constituting approximately 21 percent of the total American population [1].

An even more notable trend, and one that has particularly important consequences for future social and health policies, centers on what can be called the aging of the aging population. The number and proportion of persons 75 years of age and over will grow at a remarkable rate in the years ahead. In 1980, there were 10.0 million persons aged 75 years and over in the United States, and they constituted 4.4 percent of the total population and 39.0 percent of the overall population aged 65 and over. By 2000, there will be roughly 17.3 million persons aged 75 and over, making up 6.5 percent of the American population and 49 percent of the 65-and-over population. As with the overall older population, an even more pronounced jump in the 75-and-over group comes in the decades following the turn of the century. Thus, by 2030, the number of persons aged 75 and over is projected by the Census Bureau to be on the order of 30 million (intermediate projection); they will compose between 9 and 10 percent of the total population, although their proportion of the population 65 and over may decline slightly to approximately 47 percent [1].

Among other population data and trends, there are two particularly worthy of mention here. The first, which can largely be intuited from the figures already presented, centers on the proportion of older persons to all other younger adults. This dependency ratio is important because of what it suggests about the ratio of persons assumed to be actively participating in the labor force to those assumed no longer to be part of it. While recent inquiry has questioned elements of both these assumptions, the aggregate trends determining this ratio are nonetheless notable. Whereas in 1920 there were 7.9 aged persons for every 100 of work-force age (18–64 years old), by 1982 that ratio had risen to 18.8 to 100. By the middle of the coming century, the intermediate Census Bureau projection has the ratio at 37.8 to 100 [1]. (The short-term effect of these changes will be lessened by low birth rates and a corresponding lessening in the proportion of children who constitute the dependent-age population; however, the long-term consequences of low fertility rates are very serious, spelling as they do fewer persons entering the adult work force in coming decades.)

A last set of figures highlights a historical trend and also serves as a useful barometer of the place of older persons in the economic system. During most of the twentieth century, the proportion of older persons continuing in the active labor force has been declining. Thus, by 1975, only 14 percent of those 65 and older remained in the labor force, as contrasted with an estimated 65 to 70 percent (of the much smaller proportion) of such individuals at the beginning of the century. Since 1955, the labor force proportion of individuals aged 65 to 69 has decreased 40 percent [2]. At least until recently, there has been a presumption that declines in labor force participation of older persons would continue; however, the inflation, economic uncertainty, and concern about the future solvency of the Social Security system may be expected to alter retirement attitudes and retirement behavior as well.

ECONOMIC CONSTRAINTS AND ISSUES

Another series of factors, some related to population makeup, point toward constraints and pressures in addressing the social needs of the aged in the years ahead. These relate to economic matters, broadly construed, and the questions emerging are how long-lived and fundamental are the economic dislocations that began to emerge in the 1970s. For the elderly and programs serving the elderly, these general concerns will be directly reflected in public budget size and growth. The demographic trends discussed above point to growing demand; one must ask as well what will be the aggregate level of resources available for meeting those demands.

These broad economic issues center on the availability of natural resources (e.g., oil, minerals, water); the domestic economic alternatives center on how to stimulate real economic growth while distributing the output in what are deemed appropriate ways. The problems posed by the first set of fundamentals are potentially grave, leading as they may to shortfalls, maldistribution, and regional or worldwide violence. These issues will not fade in the near future, leaving open the clear possibility of conflict between industrial powers, between East and West, and between the industrialized countries and the less developed ones. Insofar as these matters affect the topic at hand, it can simply be noted that neither the perception nor the reality—however manufactured—of limitless natural resources will return soon, and that must color all allocation patterns worldwide. For sectors or groups such as the elderly who are seen as only marginally productive, the strains may be more pronounced.

The domestic economic debate focuses on questions of productivity and expansion. The combination of worldwide economic pressures and a conservative administration in Washington has brought to the fore an emphasis on the growth rates and supply of economic goods, placing in the shadows the issue of distribution of economic wealth and the adequacy of economic well-being among disadvantaged populations. The prevalent conservative view sees the "size of the pie" as paramount, arguing that it is near nonsense to debate how to divide up a pie that is neither large enough nor growing fast enough. The emphasis is thus placed on stimulating real economic growth, with the current interest being in stressing growth in the supply of industrial and other productive capacity rather than only in the demand for economic goods.

The liberal rejoinder is made on at least two counts. It is not at all clear that tax cuts benefiting the well-to-do will result in "supply-side" investments rather than "demand-side" indulgences. Furthermore, the notion that all parties benefit from the expanding economy predicted by conservative theorists ("a rising tide raises all ships") is questioned by liberals. The combination of minimal absolute payoff for the poor from an expanding private sector, combined with major cuts in expenditure for domestic programs yields a situation where there is (a) real loss in spending power for the poor and disadvantaged, and (b) a major injustice perpetrated as the upper quintile of American society cashes in on the new government policy.

In commenting how the elderly are affected by all this, two overall points come out. The first relates, of course, to which side of the debate one allies. To the conservative, elderly programs and benefits will be strengthened and protected by the expansion and toning up of the overall economy; in fact, the current administration, among many others, foresees major dislocations in Social Security and other old-age programs in the face of the demographic projections if economic growth does not noticeably improve. To the liberal, the poor, minority, widowed, and "old" elderly will be sacrificed on the throne of supply-side economics along with all other disadvantaged groups. Theirs is not a disagreement about the need to expand the economy—by itself, an unassailable objective—but rather their concern is on the nature of the stimulus and the identity of the beneficiaries. The current program appears likely to result in private windfalls with a residual that might also hold together existing programs for the elderly and others. But even if the "supply" aspects of the equation were only partially borne out, the liberal finds the distribution of rewards abhorrent.

This debate points out in a new light a noteworthy characteristic of the elderly population. Gerontologists have long emphasized the heterogeneity of the aged population, and nowhere is that clearer than on these macroeconomic issues. In recent decades, the elderly have proved to be slightly more conservative than the younger population, but on all but the clearest "old-age" issues (e.g., increasing Social Security benefits) the elderly are much like everyone else. Thus, one can find the well-to-do elderly in exclusive retirement communities of the Southwest, the struggling elderly of older neighborhoods throughout the East and Midwest, and the desperate elderly in the streets, haunts, and nursing homes all over the United States. There are doubtlessly many "old votes," whose concerns center on the distribution of public income and health benefits, but there is a great deal of "old money," the very existence of which reflects a greater concern with accumulation than distribution. So far, it is hard to see how a cohesive "elder position" could develop on these issues and, if it did, the direction it would take.

POLITICAL AND BUDGETARY CONSEQUENCES

At least three short-term consequences are already evidencing themselves as a result of the emergent demographics and economic issues raised here.

First and most general is the end of what might be termed "the positive-sum polity." In reference to politics, the term *positive-sum* connotes a situation where the actual or perceived benefits of a given transaction exceed the costs incurred by others. The classic instance is in pork barrel politics, where subsidies and other benefits are distributed to a number of easily identified parties, whereas the costs are ultimately borne by the central treasury and the taxpayers, who are unaware of whom their tax dollars are benefiting. More generally, the term refers to a situation where it is believed that public programs can expand in such a way that actual costs are postponed or go otherwise unrecognized. Thus, until recently, the Congress has lived largely in a positive-sum world marked, in the words of David Mayhew [3], by "credit-claiming, advertising, and position-taking."

In large part because of the demographic and economic patterns highlighted earlier, maintaining whatever positive-sum reality there ever was and perpetuating its existence are becoming increasingly difficult. The fundamental constraints have already come to have proximate consequences in the form of meaningful taxpayer constituencies, a number of initiatives to cut government spending at all levels of government, and of course, the emergence of the Reagan administration itself. What each of these occurrences does, in part, is to remove the lubricant from the positive-sum/pork barrel machine, that is, the ability to trade favors or logroll indefinitely. The fiscal pressures and tax-oriented constituencies have upset these ongoing relationships and invaded their domains.

Changes toward a more zero-sum perception of politics will constrain the array of policies for the elderly. Size alone will force more difficult decisions in the Social Security area, despite everyone's wish that the problems would somehow disappear. Smaller programs will be contained or cut as part of overall spending reduction plans or as special targets of neoconservative strategists (e.g., training monies for careers in the social services or mental health and research funds in all of the social sciences).

These changing perceptions toward the zero-sum will be most obviously manifest in the area of public spending. In the domestic arena, the big budget items are "payments to individuals" and "payments for personal health care." In 1981, the first area constituted $135 billion and the second, $52 billion. Within each of these, the totals for the elderly constituted by far the largest shares: $111 billion in income security payments to older persons (Social Security, Railroad Retirement, federal civilian employee retirement, retired military, coal miners' widows) and $37 billion for personal health care coverage (Medicare and Medicaid). The growth rates in these programs during the past decade are of greater concern than the actual amounts. The following sets of figures indicate the magnitude of the trend:

Year	Social Security (OASI)	Medicare
1966	$ 17 billion	$ 3.2 billion
1970	26 billion	5.7 billion
1975	55 billion	13.0 billion
1979	69 billion	23.0 billion
1980	100 billion	35.0 billion
1981	120 billion	40.1 billion
1982	130 billion	47.0 billion

The size and growth in other benefit programs for the elderly are great. Between 1979 and 1981 (estimated), Medicaid expenditures went from $4.3 billion to $5.5 billion, subsidized public housing jumped from $1.6 billion to $2.5 billion, and income-tested veterans' pensions rose from $3.2 billion to $3.7 billion. Even a relatively small program such as the social services provided through the Older Americans

Act has grown from a relatively insignificant $10 million in 1966 to roughly $650 million under the Title III grant program to states and communities.

Both the short-term and long-term environments will not permit this rate of growth to continue unabated. The facts that (a) upward of one-quarter of older persons are at or under 125% of the official poverty line, and (b) the rise in health care costs has been due largely to inflation and technology rather than additional benefits or beneficiaries probably will not lessen the demands for cost containment in both areas. Again, assuming only modest economic growth in the years ahead, major structural problems will emerge around 2010. With adequate lead time, there are options to be found that meet basic tenets of adequacy and equity, but the questions will have to be squarely confronted in ways that an earlier era of younger cohorts, lower inflation, and more buoyant expectations did not require.

The final observation here centers on the future political bases of old-age program formation and implementation. The old-age policy gains of past years have been prompted by a mixture of the recognition of the indisputable need of most older persons for greater support, a pervasive belief in the deservingness or legitimacy of elderly beneficiaries, the utility to politicians in being seen as supporting old-age programs, and a growing perception of the political influence the elderly might wield. However, a number of these political ingredients will fade as the developments reviewed earlier become increasingly manifest. Thus, despite data showing major income, health, housing, and other needs among a broad spectrum of the older population, the very successes that expanded programs have brought about have the potential of undermining the singular need and legitimacy that have long been associated with the elderly's claim on public benefits. In part, this is the result of data and perceptions pointing to *aggregate* improvement in the elderly's standing. But building on this is the more political aspect, namely, that other benefit groups and those concerned with overall government spending are focusing attention on the large and increasing amounts devoted to persons over 65, either as income or in-kind benefits.

As perceptions such as these are allowed to build, the political attractiveness of elderly programs begins to wane. Congress sees the costs—political as well as monetary—catching up with the benefits, and therein lies what can without exaggeration be referred to as "the new politics of aging." It is still too early to tell the extent to which economic constraints coupled with program and population growth will undermine the elderly's political standing, but it seems certain that, whatever the extent, the elderly will need to rely more on their own numbers and muscle than the evidence suggests they have to this point. The long-standing debate about "senior power" will now become more than that as the elderly find it increasingly necessary to demonstrate that they can impose electoral and other sanctions on those who fail to support their interests. Heterogeneity in the makeup of the older population will continue to impede concerted old-age political action, but there are nonetheless a core of income, health, and other programs that virtually all older persons (and younger persons, who benefit indirectly) can support. Initial skirmishes concerning Social Security indicate politicians believe that the power is there, but the real test awaits proposals for "real" cuts. The 1983 Social Security legislation suggests that it is politically possible to make cuts that were unimagined as recently as two or three years earlier.

These pressures will also affect health and other professionals serving and assisting the elderly. As budgets are constrained and volume of demand increases, criteria for service choices will need to be more rigorously clarified and eligibility for beneficiaries will be more stringent. Providers will find greater accountability pressures and more involvement by government in how their business is conducted. Among the new Washington cadre, human services "professionalism" is in low repute, and professionals in various fields serving the aged will confront this attitude as well as the more overt pressures.

The politics of aging will be more charged and program growth will be more restrained in the years ahead. A combination of pressure and leadership will have to be forthcoming in order that benefits be available to that large number of older persons requiring support. At the same time, all parties involved—public officials, professionals in the field, and the elderly themselves—need to recognize that greater targeting and efficiencies can legitimately be called for in light of the sets of constraints reviewed here.

REFERENCES

1. U.S. Bureau of the Census. *Decennial Censuses of the Population, 1900–1980 and Projections of the Population of the United States: 1982 to 2050.* Current Population Reports, Series P-25, No. 922, October 1982. Projections are Middle Series.
2. Sheppard H., Rix, S. *The Graying of Working America.* New York: Free Press, 1977.
3. Mayhew, D. *Congress: The Electoral Connection.* New Haven: Yale University Press, 1974.

Psychosocial Factors in Aging

▶

Owen W. Beard, M.D.

2

Our understanding of the problems associated with the aging musculoskeletal system will not advance until we understand those psychosocial factors that influence aging and the effects of aging. Only then can effective prophylactic steps be planned to limit those factors that can indeed be influenced and changed.

The reality of aging is apparent to all of us as we look at the people around us, but the definition of aging is a complicated and difficult task. There are variations in the way different people age, both psychologically and physiologically, as well as variations in the same person in the aging process in different systems, organs, and tissues. Adelman stated that the "biologist is not yet prepared to make a definition of biological aging" [1]. The actuarial definition is clearest but fails to satisfy the biologic scientist—that is, aging is the pattern or course of mortality as a function of age.

There is also no good agreement as to that period of life to be considered in the aging process. The biologist might say that aging begins at, or before, conception. The relationship of maternal age to the occurrence of Down's syndrome would support this consideration. The Nobel Prize winner Medawar approaches the issue of aging from the perspective of evolutionary theory and considers the entire postreproductive period as constituting aging. Since the female reproductive period is over by the end of the fifth decade and the male reproduction period may extend into and beyond the ninth decade, this definition of aging is obviously not acceptable to the majority of individuals.

DEMOGRAPHY OF AGING

In the Second Conference on the Epidemiology of Aging, sponsored by the United States Department of Health and

Human Services, there was a section on the social, psychological, and functional correlates of aging [2]. The psychosocial topics discussed and some of the points emphasized in the symposium were as follows.

Suicide

Male suicide rates are higher than female suicide rates at all ages and in all countries for which there are adequate data. Overall, the relationship between age and suicide is linear. However, aging represents a consistent factor in suicide among men but not among women.

Sex Differences in Longevity

In industrial countries, men have higher mortality rates than women, particularly for causes of death that have a behavioral component. There are suggestions that the increasing sex differentials in longevity in the United States may be reversed in the future.

Mortality After Early Retirement

Mortality after early retirement is higher than expected in a comparable-age working population. However, analysis of the factors involved reveals that survival after such early retirement is related to the previous health status in the two groups and not to the early retirement per se.

Normal retirement is not obviously detrimental to survival. Since most men retiring at age 65 are in relatively good health, one would not expect to see an immediate rise in mortality. In fact, the honeymoon-phase theory suggests that the initial period after retirement is not stress evoking. But if there is a honeymoon phase after normal retirement, there should logically follow a disenchantment phase. There is some evidence to suggest that there may be a disenchantment phase.

1. Martin and Doran [3] found elevations in illness rates 4 to 6 years after retirement among blue-collar workers.
2. Solem [4] reported elevated mortality during the third year after compulsory retirement at age 70 in Norway.
3. Stokes and Maddox [5] have also shown that blue-collar workers experience a substantial decline in satisfaction 3 to 5 years after retirement.

Involuntary Relocation

The effects of involuntary relocation on the health and behavior of the elderly remain controversial. A consistent and coherent picture does not emerge from perusal of the published studies.

LONGEVITY

Consideration of the actuarial definition of aging—the pattern or course of mortality as a function of age—suggests that it might be informative to look at some of the psychosocial factors as predictors of longevity. Even though a clear, valid, and reliable definition of aging remains to be formulated [6], whatever the process may be, it is only in people's temporary avoidance of mortality that aging has an opportunity to manifest itself.

Diet

Too much or too little food reduces longevity. In the First Duke Longitudinal Study [7], it was found that both the tenth percentile with the lowest weight-height ratio and the tenth with the highest weight-height ratio had more illnesses and less longevity than the middle 80 percent. However, no one has satisfactorily separated the effects of obesity and lack of exercise.

Exercise

Several studies have found that older persons who are more active and get more exercise tend to live longer than those getting less exercise. However, decrease in exercise tends to accompany obesity and chronic illness, and there is the possibility that greater amounts of exercise may just be an indicator of better general health, which leads to the greater longevity.

Smoking

The association between cigarette smoking and higher mortality is now well established.

Retirement and Work

Retired persons have a higher mortality than those who continue to be gainfully employed. However, this is almost entirely due to the fact that only the more healthy continue to work, while the sicker individuals retire.

Marital Status

Among the aged, married persons have lower mortality rates than any of the nonmarried categories.

Greater Social Activity

Greater social activity is modestly correlated with greater longevity. Berkman [8] found in her study of the habits of 7,000 people that those who took an active role in socializing were more likely to live longer. Those who stayed home were more likely to be overweight and to drink and smoke more. However, is it greater social activity that leads to better health and longevity *or* better health that leads to increased social activity?

Predictors of Longevity
There are some predictors of longevity that in our current state of knowledge are less subject to modification than those just discussed.

Heredity Many studies have shown a general association between the longevity of parents and offspring. However, we do not know how much of the association, if any, is due to genetic heredity and how much is due to the inheritance of social, psychological, and economic environments similar to those of the parents.

Sex Women live longer, but how much of this difference is due to sex-linked genetic differences and how much is due to differences in life-style between men and women are not known.

Race Blacks have higher mortality than whites at most ages.

Intelligence When other variables are controlled, intelligence is not a significant predictor of longevity.

Socioeconomic Status Generally there is a positive relationship between socioeconomic status and longevity, but the why of the relationship is not clear.

Conclusions
There are several behavioral and social factors, such as diet, exercise, smoking, and marriage, that are predictors of longevity, as are several social characteristics less subject to modification, such as sex, race, intelligence, and socioeconomic status. However, the reasons or explanations why these variables predict longevity are usually ambiguous at best.

AGING CHANGES IN SPECIAL SENSES

Some effects of aging on people's special senses have potential implications as psychosocial stress factors leading to musculoskeletal disease. Through an understanding of these factors it might be possible for a society to make the appropriate environmental changes that would minimize their adverse effects on the aged.

Accident injuries become more frequent and more serious in later life; therefore, it is important that we look especially at the causative factors that might be ameliorated by prophylactic measures. Accidents are the fifth most common cause of death among older persons. The aged, who constitute about 11 percent of the national population, suffer about 23 percent of all accidental deaths [9].

Several factors make the elderly person more vulnerable to accidental injury.

Various Diseases
Various diseases such as arthritis and neurologic disorders can impair coordination and balance.

Drugs and Alcohol
Many drugs and alcohol can result in decreased cognitive function, impairment of judgment, and drowsiness.

Mental Depression
Mental depression or preoccupation with personal problems can lead to decreased awareness of environmental hazards.

Diminished Hearing
Diminished hearing with aging may decrease awareness of potential hazards. With aging there is a progressive loss in hearing, first for high frequencies but, with progression, for the mid and lower frequencies as well. Most of the warning signals in our environment are high pitched (e.g., horns, bells, sirens, telephones). It is important that correctable hearing problems in the aged be treated and that prosthetic aids be utilized where appropriate. Such attention to these communication problems will not only make the aged person more aware of the potential hazards in the environment but will also be a positive factor in contributing to the person's interaction and enjoyment of the environment.

Decrease in Tactile Sensation
With aging there is a decrease in tactile sensation with a decrease in the ability to distinguish degrees of heat and cold. Burns are especially disabling to the elderly, since the recovery process is so slow. Among the many environmental changes that should be considered in an effort to minimize burns in the elderly is to use lower temperature settings on water-heater thermostats.

Diminished Eyesight
Diminished eyesight with aging is one of the most important factors in leading to accidents.

Glare Glare from uncontrolled natural light and from unbalanced artificial light sources is the single greatest difficulty encountered by the aged eye. The environment suitable for the elderly should be well lighted but with surfaces that do not increase glare.

Color The elderly eye has problems distinguishing between pastel shades of color. All colors tend to appear faded, with red fading the least and the cool colors such as green and blue fading the most. When two intense colors such as red and green meet, their boundary appears unstable and seems to overlap, thus presenting problems in recognition of hazards.

Depth Perception Depth perception is impaired in the elderly, so that if both stairs and risers have a floral-design carpet, or are the same solid color, the aged may have problems with falling on stairs. Falls are a common cause of fatality among the aged. Some precautionary measures that might be taken include

1. Properly lighted stairways and stairways with handrails
2. Carpeting that is fixed to the floor
3. Elimination of throw rugs
4. Use of a floor wax that provides good traction
5. Furniture arranged so that there are no obstacles
6. Well-lighted bathrooms equipped with grab bars
7. Bedside nightlights

Traffic Accidents Traffic accidents are the most common cause of accidental death in the 65- to 74-year age range. The elderly person's ability to drive safely may be impaired because of the increased sensitivity to glare and poor adaptation to the dark. However, nonvisual problems such as slower reaction time, decreased coordination, and poor mental or physical health in general also contribute to risk of traffic accidents.

Are older individuals safe or unsafe drivers as a group? Conflicting answers have been given to this question. Older drivers have a relatively small number of crashes, but they drive less, so that when miles driven or exposure is taken into consideration, older drivers have a higher crash rate than any group except the under 25-year age group [10]. In a study of the factors responsible for 56 crashes in a three-year period for 354 male and female white drivers, no association was found between vision, hearing, cardiovascular status, and general functional status. In spite of the above conclusion, the predominant driver error recorded as responsible for these crashes was failure to stop at a traffic signal or stop sign or failure to yield at an intersection. Pertinent to the interpretation of these data is the fact that road signs are designed for people with 20/30 or 20/40 vision. Only one driver was charged with speeding and two were noted to be under the influence of alcohol [11].

Decrease in Sense of Smell
Decrease in the sense of smell with aging also leads to impairment of the ability to recognize smoke early. The relationship of the decrease in smell, with a consequent decrease in taste for and interest in food, to malnutrition in the aged is an

area that needs more research. The potential role of malnutrition in the demineralization of aged bones is still an unsettled area.

SUMMARY

A satisfactory definition of the aging process remains to be formulated. Thus, it is not clear how psychosocial factors may either directly or indirectly affect aging. Since, in general, aging is related to time or duration of life, this chapter has reviewed those psychosocial factors that have an effect on the pattern or course of mortality as a function of age.

REFERENCES

1. Adelman, R.C. Definition of biological aging. In Haynes, S.G., Feinleib, M. (eds.), *Second Conference on the Epidemiology of Aging.* NIH Publication No. 80-969. Washington, D.C.: U.S. Dept. of Health and Human Services, Public Health Service, National Institutes of Health, 1980, pp. 9–13.
2. Social, psychological and functional correlates of aging. In Haynes, S.G., Feinleib, M. (eds.), *Second Conference on the Epidemiology of Aging.* NIH Publication No. 80-969. Washington, D.C.: U.S. Dept. of Health and Human Services, Public Health Service, National Institutes of Health, 1980, pp. 139–285.
3. Martin, J., Doran, A. Evidence concerning the relationship between health and retirement. *Sociol. Rev.* 14:329–343, 1966.
4. Solem, P.E. Paid work after retirement age, and mortality. In *Retirement: Norwegian Experiences.* Oslo, Norway: The Norwegian Institute of Gerontology, 1976, pp. 33–65.
5. Stokes, R.G., Maddox, G.L. Some social factors on retirement adaptation. *J. Gerontol.* 22:329–333, 1967.
6. Ostfeld, A.M., Gibson, D.C. (eds.). *Epidemiology of Aging.* DHEW Publication No. (NIH) 75-711. Washington, D.C.: U.S. Dept. of Health, Education, and Welfare, Public Health Service, National Institutes of Health, 1975.
7. Palmore, E. Health practices and illness among the aged. *Gerontologist* 10:313–316, 1970.
8. Berkman, L.F., Syme, S.L. Social networks, lost resistance, and mortality: A nine-year followup study of Alameda County residents. *Am. J. Epidemiol.* 109:186–204, 1979.
9. Accidents and the elderly. In Warfel, B.L. (ed.), *Information on Aging.* Detroit, Mich.: Wayne State University Institute of Gerontology of the University of Michigan, July 1981, pp. 7–8.
10. Hogue, C.C. Epidemiology of injury in older age. In Haynes, S.G., Feinleib, M. (eds.), *Second Conference on the Epidemiology of Aging.* NIH Publication No. 80-969. Washington, D.C.: U.S. Dept. of Health and Human Services, Public Health Service, National Institutes of Health, 1980, pp. 127–135.
11. Palmore, E. *Normal Aging II.* Durham, N.C.: Duke University Press, 1974.

The Biology and Physiology of Aging

▶

Edward J. Masoro, Ph.D.

3

The ideal way to start any discussion of the biology of aging would be to define aging. Although all of us are familiar with aging from observing our family, friends, and pets, an attempt at defining aging on a rigorous biologic basis reveals that the necessary knowledge base does not exist. At this time, therefore, a description of the major characteristics of mammalian aging provides the best starting point.

The following are the salient characteristics of aging in mammals: (a) the probability of adults dying increases with increasing age, (b) body composition changes with age, (c) physiologic deterioration occurs with advancing age, and (d) diseases have increasingly serious consequences with advancing age, and certain diseases occur in an age-associated manner. A consideration of each of these characteristics offers a basic picture of the aging process consistent with the current state of gerontologic knowledge.

CHARACTERISTICS OF AGING

Mortality

Survival curves are a particularly informative way to view mortality. For example, consider the theoretical curves [1] presented in Figure 3-1. The y-axis depicts percent survivors and the x-axis, age of the population. Curve *a* is an exponential decay curve, a type of survival curve not seen with mammalian species; survival curves of mammals are shifted to the right of curve *a*. Curve *b* is one that might be expected for small mammals (e.g., rats) living in the wild, while curves *c* and *d* might be seen as the animals are maintained in increasingly protected environments. For example, curve *c* could well be seen with laboratory rats housed in conventional facilities, and curve *d* would occur with laboratory rats housed in a barrier facility that provides a specific pathogen-free state. Curve *d* has a rather rectangular shape, and the progressive change in shape between curves *b* and *d* is often referred to as the rectangularization of the survival curve. It is important to note that the maximum life span for a particular species is the same for curves *b, c,* and *d*. The rectangularization of the survival curves reflects the fact that the length of life of a greater percentage of the population of the species approaches the maximum life span (which for rats is approximately 3 years). Thus, for mammals living in protected environments, mortality is clearly associated with advanced age.

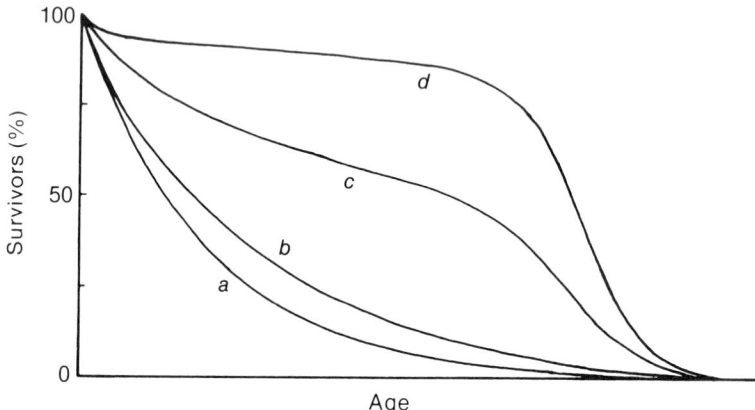

Figure 3-1

Survival curves for four hypothetical populations. Curve *a* is an exponential decay curve; curve *b* is typical of many wild animal populations; curves *c* and *d* are curves obtained with animal or human populations protected by technologic advances (e.g., sanitation engineering) and by medical progress.
SOURCE: Reproduced with permission from Kirkwood and Holliday [1].

Survival curves for humans [2] are depicted in Figure 3-2. The curve on the left is that for American men during the mid-1970s. In comparison, the curve for men in Ancient Roman times is similar to that of curve *b* in Figure 3-1. Over the centuries, the development of technology and medical science has rectangularized the human survival curve, but the maximum life span—approximately 100 years—remains unchanged. Future developments in technology and medicine are expected to rectangularize the survival curve further, resulting in curves such as the center one in Figure 3-2. It is also possible that research aimed at modulating the aging process may result in extending the maximum life span and, if so, might well result in survival curves such as the one on the right in Figure 3-2.

Body Composition

A typical picture of the changes in human body composition associated with increasing age [3] is presented in Figure 3-3.

The Biology and Physiology of Aging

Although many men and women show an increase in body weight between 45 and 60 years of age, in the case of this particular man total body weight remained rather constant during these years. However, as is the case in almost all people, his lean body mass fell progressively with age (loss in lean body mass starts at about 30 years of age). Since the

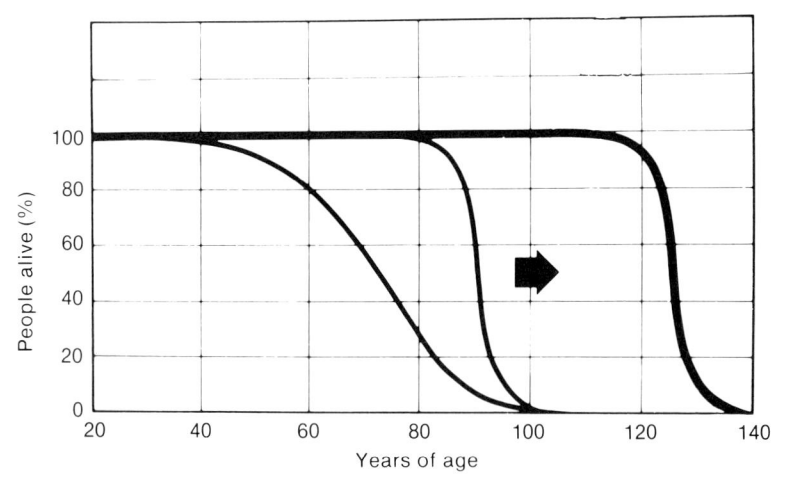

Figure 3-2

Human survival curves. The curve on the left was calculated from the 1974 life expectancy compilations for the American man by the U.S. Public Health Service. The center curve is a projection of the possible effect of continued advances in technology and medicine on the survival curve of the American male. The curve on the right is a projection of the possible effect of the applications of aging research on the survival of the American male.
SOURCE: Modified from Robinson [2].

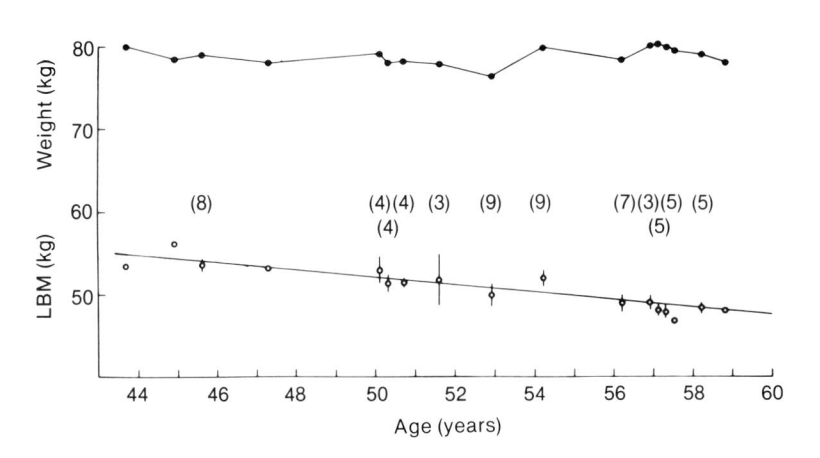

Figure 3-3

Change in lean body mass (LBM) and body weight as a function of age in one human subject. Number of measurements is in parentheses, and vertical bars are the standard deviation.
SOURCE: Reproduced with permission from Forbes [3].

difference between body weight and lean body mass is the fat mass, Figure 3-3 shows that fat mass progressively increased between 45 and 60 years of age. Indeed, in almost all men and women living in developed nations, fat mass increases with age at least until late middle age [4].

It has long been postulated that adiposity causes a decrease in longevity. The basis of this belief derives from extensive data on insured persons compiled and analyzed by the Society of Actuaries [5]. These data, summarized in Figure 3-4, show a linear relationship between overweight (per-

Figure 3-4

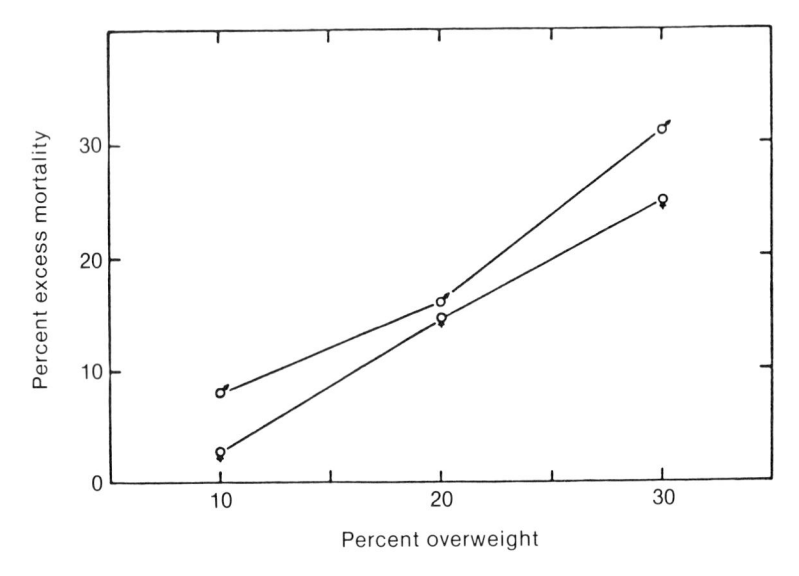

Effect of overweight on excess mortality in men (♂) and woman (♀) between 40 and 69 years of age. Data from the Society of Actuaries Build and Blood Pressure Study, 1959.
SOURCE: Reproduced with permission from Andres [6].

centage over average weight of the insured population) and excess mortality. However, analysis of a great many recent studies by Andres [6] indicates that adiposity over a moderate range does not influence the mortality rate; a typical example is his analysis of the Framingham Study, shown in Figure 3-5. Two possible reasons are apparent for the striking difference between the conclusion drawn by Andres and that classically held. First, those persons who are insured may represent a biased sample of the nation's population. Second, for the point of reference in his analysis, Andres used the "ideal body weight" figures (for weight and sex) published in the *Statistical Bulletin* in 1959 [7], while the Society of Actuaries used the average weight of the insured population as the point of reference. The latter weight is considerably

The Biology and Physiology of Aging

greater than the former. Those practitioners using the data of the Society of Actuaries have, in most cases, mistakenly assumed that the point of reference was ideal body weight when it was, in fact, the average weight of the insured population.

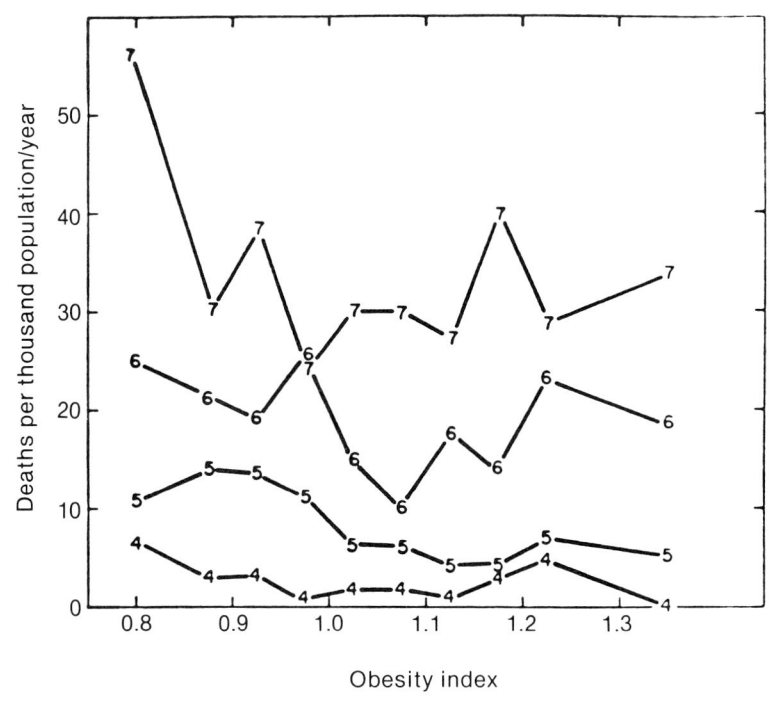

Figure 3-5

Relation of obesity to mortality in men: The Framingham Study. 4 = 35–44 years old; 5 = 45–54 years old; 6 = 55–64 years old; 7 = 65–75 years old.
SOURCE: Reproduced with permission from Andres [6].

Physiologic Deterioration

There is a large data base that establishes physiologic deterioration as a characteristic of aging [8]. The essence of these data is summarized in a graph (Fig. 3-6) by Shock [9]. These physiologic changes lead to a loss in capacity to carry out activity and a loss in ability to meet challenges.

An example of the loss in capacity to carry out activity is the decline with increasing age in maximal oxygen uptake ($\dot{V}O_2$max) per unit of body weight in response to exercise in men, shown in Figure 3-7. Dehn and Bruce [10] feel that this age-related decline in the intensity of exercise that can be performed relates primarily to the progressive deterioration in cardiovascular function.

An example of the reduced ability to meet challenges is

provided by the study by Shock [9], which tested the ability of young and old people to respond to an acid load (Fig. 3-8). The ingestion of 10 g of NH_4Cl (an acidifying agent) caused a much more marked fall in blood pH and one that lasted much longer in old people than in young people. It was

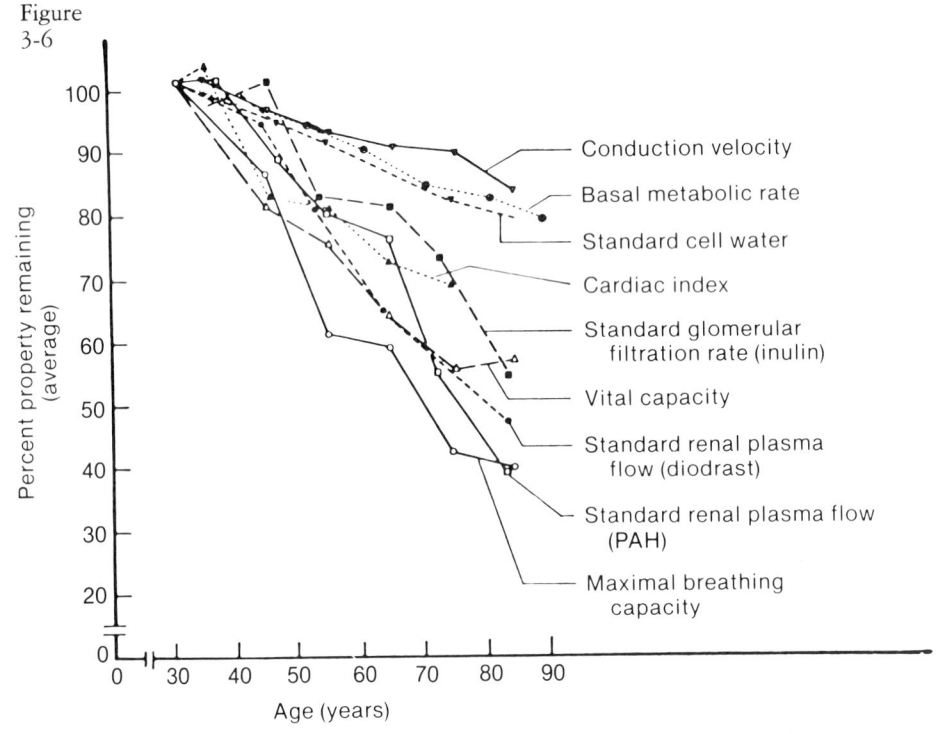

Figure 3-6

[Graph showing percent property remaining (average) on y-axis from 0 to 100, and Age (years) on x-axis from 0 to 90, with curves labeled: Conduction velocity, Basal metabolic rate, Standard cell water, Cardiac index, Standard glomerular filtration rate (inulin), Vital capacity, Standard renal plasma flow (diodrast), Standard renal plasma flow (PAH), Maximal breathing capacity]

Age changes in physiologic function based on data from cross-sectional studies. Data are expressed as percent of mean value at age 30 years.
SOURCE: Reproduced with permission from Shock [9].

further shown that the loss in ability to meet this acid-base challenge related to declining kidney function [11].

Disease

Although it has been suggested that aging is the result of disease, this concept is not generally accepted. However, it is recognized that aging and disease are related [12]. Indeed, just as NH_4Cl administration has more deleterious consequences in the old than in the young, so do disease processes (e.g., the effects of infectious diseases such as influenza). Moreover, there do exist age-related or age-associated diseases [13] (e.g., atherosclerosis, cerebrovascular accidents, carcinoma, osteoporosis, osteoarthritis, diabetes mellitus,

The Biology and Physiology of Aging

senile dementia of the Alzheimer type), which either appear at advanced ages or become debilitating in old age. However, even in these diseases, it seems likely that aging provides the conditions necessary for their development rather than the diseases causing aging.

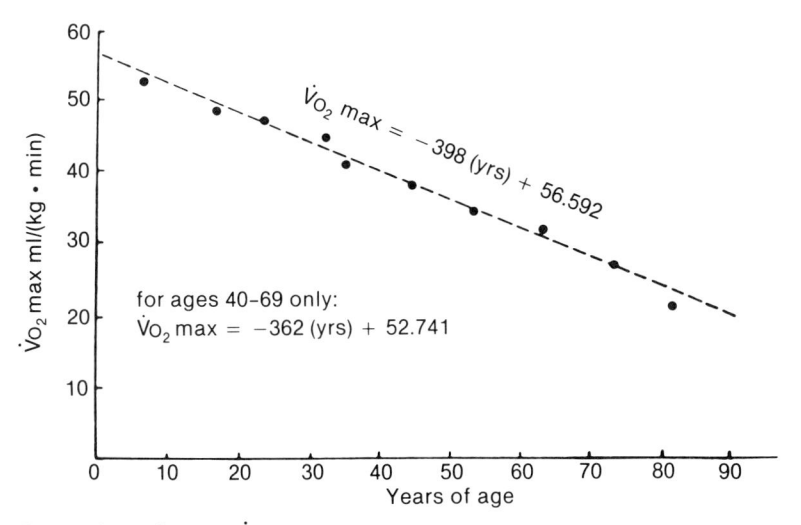

Figure 3-7

Regression of mean $\dot{V}O_2$max per decade of age of 700 observations in healthy boys and men recalculated from 17 studies in the literature. Unlike more conventional plots of $\dot{V}O_2$max in liters per minute against age, which show an increase with growth and development in childhood and a progressive decline after adolescence, when $\dot{V}O_2$max is corrected for body weight and expressed in ml/kg · min, there is a fairly uniform decrease throughout life.
SOURCE: Modified with permission from Dehn and Bruce [10].

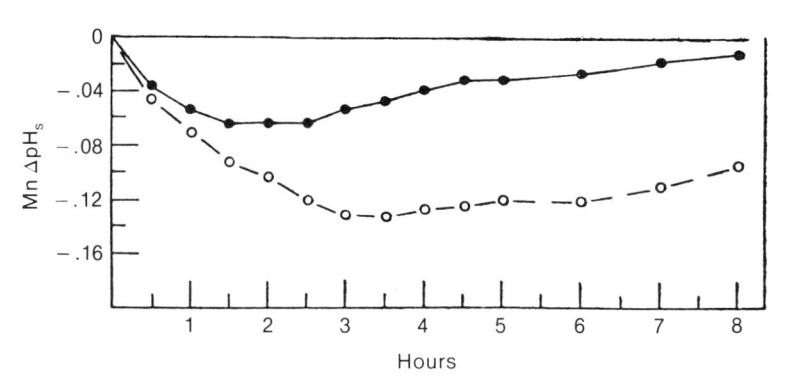

Figure 3-8

Response of blood pH to an ingested acid load (10 g NH_4Cl). Solid circles denote young people, and open circles denote old people.
SOURCE: Modified with permission from Shock [9].

BIOLOGIC NATURE OF AGING

The characteristics of aging described above are reflections or manifestations of aging, but they are not likely to be the primary events of this process. At this time, the nature of the primary aging process is not known. However, it is likely to involve a progressive change in the fundamental function or functions of all cells or key cells. In regard to key cells, for example, it can be theorized that altered functioning of neurons of the hypothalamus in mammals could result in widespread physiologic alterations with characteristics like those seen in aging. The challenge to the biologic gerontologist of today is to learn in which cells the primary aging processes occur and which cell processes are involved.

EXPERIMENTAL APPROACHES TO THE EXPLORATION OF BASIC AGING PROCESSES

In searching for the nature of the aging process, a wide range of research areas have been investigated in recent years. Although the experimental measurements varied greatly from study to study, all involved one or more of the following general approaches: (a) the testing of a theory of aging, (b) the exploration of a manifestation of aging at the cellular or molecular level, (c) the analysis of differences among species of markedly different life spans, and (d) the analysis of the actions of manipulations that either decrease or increase life span. None of these has as yet provided a breakthrough. A combination of these approaches will undoubtedly be required to begin to unravel this problem. For instance, the extension of life span by food restriction viewed in the light of the other three approaches appears to be a promising area for further study.

RETARDATION OF AGING BY FOOD RESTRICTION

In the 1930s, McCay and colleagues [14] showed that food restriction severe enough to retard growth markedly, resulted in an increased length of life in rats. Using different modes of restricting food, this finding has been confirmed for a variety of laboratory rodents [15]. Indeed, Ross [16] has shown that with rats offered a free choice of three food mixtures and allowed to eat ad libitum, there is an inverse relationship between the total amount of food the rats ate and the length of life (Fig. 3-9).

Recent research in our laboratory and those of others strongly indicates that food restriction prolongs life by delaying the aging process [17]. A brief description of our work as well as the key findings of other investigators should provide an overview of the basic evidence in support of this thesis.

In our research we used specific pathogen-free Fischer 344 male rats. At 6 weeks of age, the rats were divided into

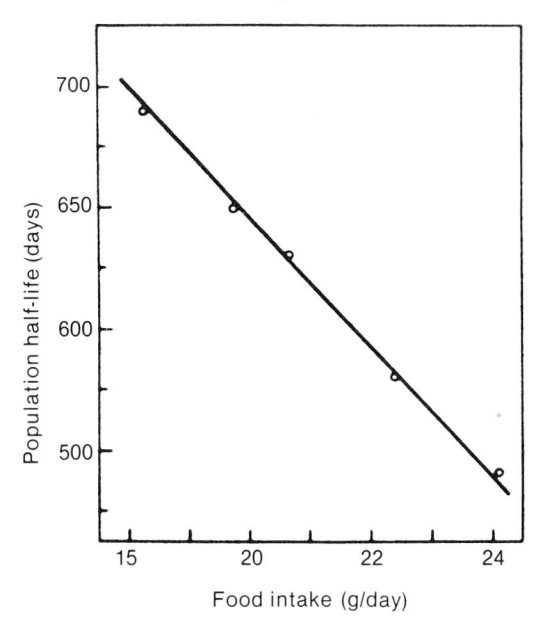

Figure 3-9

Relationship between length of life and quantity of food consumed by male rats given freedom of dietary choice.
SOURCE: Reproduced with permission from Ross [16].

two groups, group A and group R. Group A rats were fed ad libitum, and group R rats were given 60 percent of the mean ad libitum intake. Group A and group R were each also divided into three subgroups: one for the study of longevity, another for the longitudinal study of body composition, and a third for the cross-sectional study of biochemical and physiologic parameters. The growth of the two rat groups [18] is shown in Figure 3-10.

The survival curves, shown in Figure 3-11, summarize the longevity findings [18]. The striking feature is that food restriction shifted the curve to the right but made it less rectangular. These data make it clear that food restriction increased the maximum length of life; indeed, when the last rat of group A died, more than 60 percent of the group R rats were still living. The effect of food restriction on longevity should be contrasted with the effects of technology and medicine; the latter two factors increased the mean length of life, but they did not influence the maximum length of life.

Our longitudinal study of body composition showed that fat mass increased during the adult life of both group A and group R rats until about 70 percent of the life span was reached, after which it declined with advancing age [19]. Group R rats had a smaller total fat mass than group A rats,

Figure 3-10

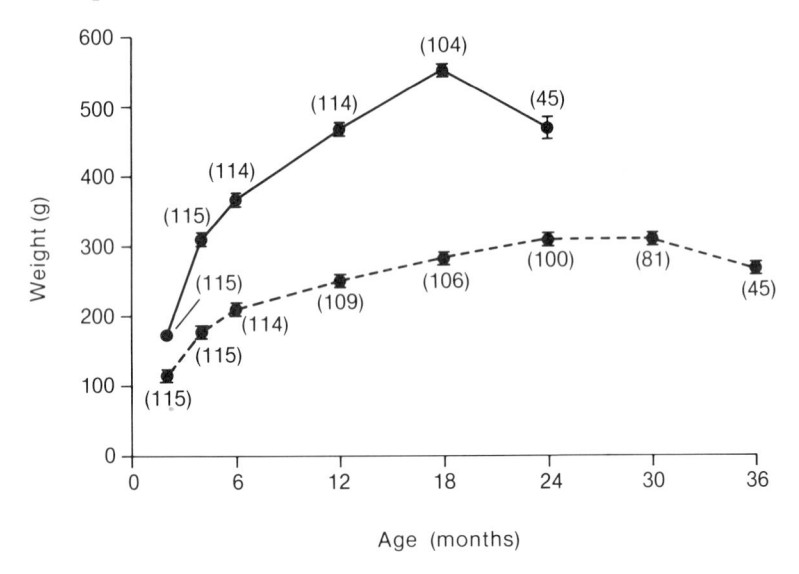

Changes in body weight during life span of Fischer 344 rats. The solid line refers to group A rats and the broken line to group R rats. The data are mean values ± S.E. for all animals alive at a given chronologic age. The numbers in parentheses are the number of animals so analyzed.
SOURCE: Reproduced with permission from Yu et al. [18].

Figure 3-11

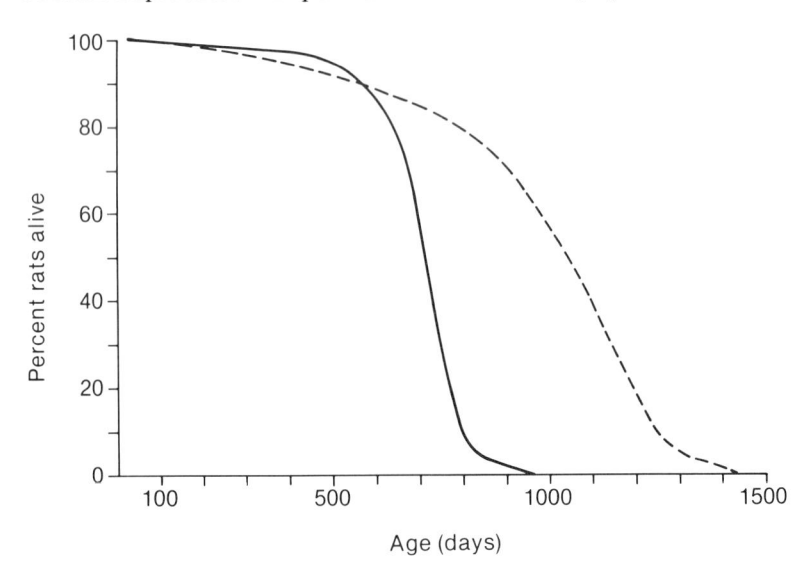

Survival curves for group A and group R rats. The solid line refers to group A rats ($n = 115$) and the broken line to group R rats ($n = 115$).
SOURCE: Reproduced with permission from Yu et al. [18].

even when the data are expressed in terms of percent of body mass. In group A rats, there was no correlation between fat mass and length of life, but in group R rats there was a significant positive correlation between maximum attained fat mass and length of life (Table 3-1). Thus, it seems unlikely that food restriction promotes longevity by reducing fat mass.

Relationship Between Fat Mass and Length of Life in Group A and Group R Rats

Table 3-1

Group	Parameter	r	P
A	Fat, % body weight at 8 months of age	−0.12	n.s.
	Fat, maximum mass attained	−0.04	n.s.
	Fat, maximum % body weight attained	−0.10	n.s.
R	Fat, % body weight at 8 months of age	0.12	n.s.
	Fat, maximum mass attained	0.63	< 0.001
	Fat, maximum % body weight attained	0.63	< 0.001

n.s. = not significant.
SOURCE: Data from Bertrand et al. [19].

Food restriction modulates age-related changes in the physiologic system. A most striking example is the extent to which it prevents the age-related decline [20] in the response of adipocytes to the lipolytic action of glucagon (Fig. 3-12). In group A rats, there was a marked loss of this responsiveness in the isolated adipocyte between 6 and 15 weeks of age and a total loss by 6 months of age. In contrast, in the case of adipocytes from group R rats, no loss in the lipolytic response to glucagon was observed until after 12 months of age. Furthermore, at no age was the response to glucagon totally lost in adipocytes from group R rats; for example, adipocytes from 36-month-old group R rats still showed a significant response. Another example is seen in the age-related increases in serum cholesterol that occur in rats, just as they do in humans [21]. Although food restriction did not influence the serum cholesterol level in 6-month-old rats, it markedly delayed the age-related increase in concentration of this lipid (Fig. 3-13); it was not until 30 months of age that group R rats showed serum cholesterol levels that are achieved as early as 18 months of age in group A rats. These are not isolated examples of food restriction's delaying or preventing age-related physiologic change. Food restriction has been shown to prevent or delay the following age-related changes in physiologic activity: loss of responsiveness to hormones [20, 22], alterations in lipid metabolism [22], alterations in adipose tissue morphology and function [19], alterations in skeletal muscle structure and function [23],

Figure 3-12

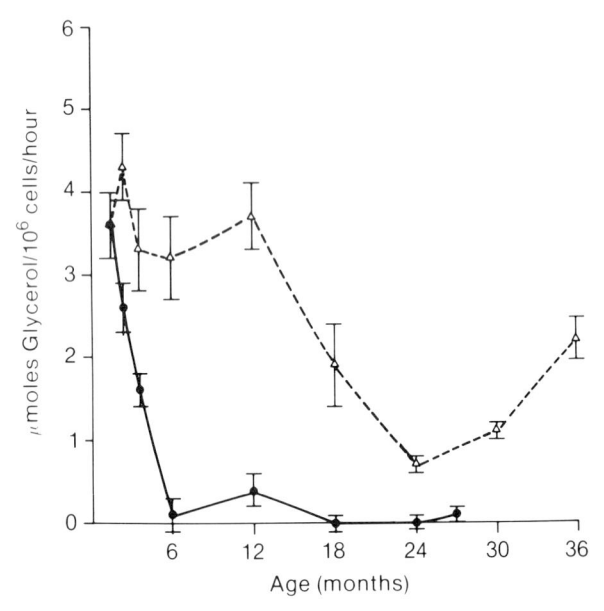

Glucagon-promoted lipolysis in adipocytes from group A (●——●) and group R (△------△) rats: the effects of age on the response of isolated adipocytes to the lipolytic action of glucagon (1 μg/ml).
SOURCE: Reproduced with permission from Masoro et al. [17].

alterations in smooth muscle functions [24], and changes in immune function [25].

Food restriction also delays age-related disease [16]. An example of this is chronic nephropathy in group A and group R rats [18], as reported in Table 3-2; grade 0 indicates no lesion, with increasingly severe lesions occurring progressively in order of grades 1, 2, 3, 4, and 5. It is clear that these lesions rapidly increased in severity with age in group A rats, while group R rats showed a much slower rate of progression in severity. Indeed, at death almost all group A rats showed severe renal lesions, but few group R rats had such lesions. This effect of food restriction on age-related disease is not restricted to renal lesions but has been seen with many, though not all, types of tumors [16] and with such problems as cardiac fibrosis and bile duct hyperplasia [18].

It seems clear from these data that food restriction does delay the aging process. The important question to consider at this time is what is the mechanism by which food restriction effects the antiaging action? In 1935, McCay et al. [14] theorized that food restriction delayed aging by delaying development or prolonging immaturity. This is probably not the major mechanism, however. Recent findings show that food restriction prolongs life in rats when started at 12

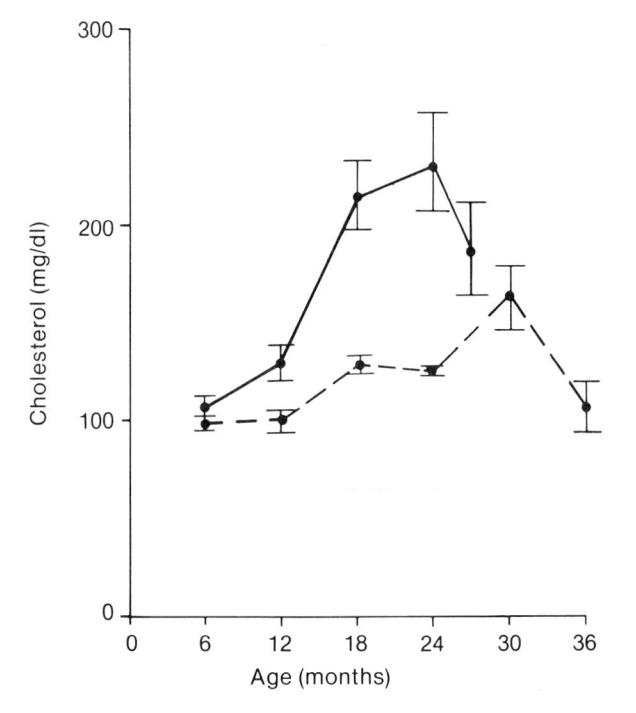

Figure 3-13

Age and postabsorptive serum total cholesterol levels of group A (solid line) and group R (broken line) rats.
SOURCE: Reproduced with permission from Liepa et al. [21].

months of age [26] and in mice when started after full growth has been attained [27]. These studies also rule out the suggestion that food restriction prolongs life by reducing the rate of growth or prolonging the duration of growth. In addition, our work, as reported above, rules out the possibility that food restriction prolongs life by reducing the adipose mass.

An interesting hypothesis [28] is that food restriction prolongs life by reducing fuel flux and thereby the rate of electron transport in the mitochondria. It could be further postulated that damaging substances such as superoxide and hydroxyl radicals generated during electron transport [29] are responsible for aging. The extent of their damaging action would be expected to relate to their rate of production and to the level of enzymatic machinery available (e.g., superoxide dismutase) that destroys these compounds. However, unpublished work from our laboratory makes this hypothesis unlikely, since it was found that the rate of fuel use per gram of metabolizing tissue is not decreased by food restriction.

While it has been clearly established that food restriction prolongs life by delaying aging, the mechanism by which this is achieved is totally unknown. Data from recent studies have made all of the most likely of previously proposed hypotheses unlikely.

Table 3-2 Chronic Nephropathy in Sacrificed Rats

Rat Group	Age Range (months)	\multicolumn{6}{c}{Grade (number of rats with lesion)}
		0	1	2	3	4	5
A	5½–6½	4	5	0	0	0	0
A	11½–12½	0	1	8	1	0	0
A	17½–18½	0	0	0	9	1	0
A	23½–24½	0	0	0	2	8	0
A	26½–27½	0	0	0	2	2	4
R	5½–6½	8	2	0	0	0	0
R	11½–12½	3	7	0	0	0	0
R	17½–18½	0	3	7	0	0	0
R	23½–24½	0	0	10	0	0	0
R	29½–30½	0	0	8	1	0	0
R	35½–36½	0	0	2	8	0	0

SOURCE: Modified from Yu et al. [18].

CONCLUSIONS

Although many characteristics of aging are well known, the nature of the primary aging process or processes has yet to be discovered. However, the secondary processes or manifestations of the primary process are of concern in their own right because of their medical importance. They encompass the increasing probability of death with increasing age, the change in body composition (e.g., decreasing bone and muscle mass) with increasing age, and the age-related deterioration in most physiologic systems. Moreover, there are complex interactions between aging and disease. These include the increasingly detrimental effects with advanced age of the commonly occurring diseases (e.g., influenza), which probably result from the reduced ability to cope with challenges successfully because of the deterioration in the physiologic systems. Also, there are diseases that are age-related in terms of their time of occurrence during the life span (e.g., senile dementia of the Alzheimer type); these diseases probably require the occurrence of the primary aging process as a substratum for their manifestation.

In rodents, food restriction markedly extends the length of life; it does so by slowing the rate of aging. Thus, it is not surprising to find that food restriction modulates age-related physiologic deterioration and age-related disease. Moreover, our data suggest that such nutritional manipulations hold great promise as an approach for the experimental exploration of the primary aging process. It also seems probable that nutrition may be a practical way to modulate aging effectively in all mammals, including humans.

REFERENCES

1. Kirkwood, T.B.L., Holliday, R. The evaluation of ageing and longevity. *Proc. R. Soc. Lond. [Biol.]* 205:531–546, 1979.
2. Robinson, A.B. Molecular clocks, molecular profiles and optimum diets: three approaches to the problems of ageing. *Mech. Ageing Dev.* 9:225–236, 1979.
3. Forbes, G.B. The adult decline in lean body mass. *Hum. Biol.* 48:161–163, 1976.
4. Malina, R.M. Quantification of fat, muscle and bone in man. *Clin. Orthop.* 65:9–38, 1969.
5. Society of Actuaries. *Build and Blood Pressure Study, 1959*, vol. 1. Chicago: Society of Actuaries, 1960, pp. 1–268.
6. Andres, R. Influence of obesity on longevity in the aged. *Adv. Pathobiol.* 7:238–246, 1978.
7. New Weight Standards for Men and Women. *Stat. Bull. Metropol. Life Ins. Co.* 40:1–3, 1959.
8. Masoro, E.J., Bertrand, H., Liepa, G., Yu, B.P. Analysis and exploration of age-related changes in mammalial structure and function. *Fed. Proc.* 38:1956–1961, 1979.
9. Shock, N.W. The science of gerontology. In Jeffers, E.C. (ed.), *Proceedings of Seminars 1959–61, Durham, N.C., Council on Gerontology.* Durham, N.C.: Duke University Press, 1962.
10. Dehn, M.M., Bruce, R.A. Longitudinal variations in maximal oxygen intake with age and activity. *J. Appl. Physiol.* 33:850–857, 1972.
11. Adler, S., Lindeman, R.D., Yiengst, M.J., et al. Effect of acute acid loading on urinary acid excretion by aging human kidney. *J. Lab. Clin. Med.* 72:278–289, 1968.
12. Upton, A.C. Pathobiology. In Finch, C.E., Hayflick, L. (eds.), *Handbook of Biology of Aging*. New York: Van Nostrand Reinhold, 1977, pp. 513–535.
13. Martin, G.M. Genetic and evolutionary aspects of aging. *Fed. Proc.* 38:1962–1967, 1979.
14. McCay, C.M., Crowell, M.F., Maynard, L.A. The effect of retarded growth upon the length of life span and upon the ultimate body size. *J. Nutr.* 10:63–79, 1935.
15. Young, V.R. Diet as a modulator of aging and longevity. *Fed. Proc.* 38:1994–2000, 1979.
16. Ross, M.H. Nutrition and longevity in experimental animals. In

Winick, M. (ed.), *Nutrition and Aging.* New York: Wiley, 1976, pp. 43–47.
17. Masoro, E.J., Yu, B.P., Bertrand, H.A., Lynd, F.T. Nutritional probe of the aging process. *Fed. Proc.* 39:3178–3182, 1980.
18. Yu, B.P., Masoro, E.J., Murata, I., et al. Life span study of SPF Fischer 344 male rats fed ad libitum or restricted diets: longevity, growth, lean body mass and disease. *J. Gerontol.* 37:130–141, 1982.
19. Bertrand, H.A., Lynd, F.T., Masoro, E.J., Yu, B.P. Changes in adipose mass and cellularity through the adult life of rats fed ad libitum or a life-prolonging restricted diet. *J. Gerontol.* 35:827–835, 1980.
20. Bertrand, H.A., Masoro, E.J., and Yu, B.P. Maintenance of glucagon-promoted lipolysis in adipocytes by food restriction. *Endocrinology* 107:591–595, 1980.
21. Liepa, G.U., Masoro, E.J., Bertrand, H.A., Yu, B.P. Food restriction as a modulator of age-related changes in serum lipids. *Am. J. Physiol.* 238:E253–E257, 1980.
22. Yu, B.P., Bertrand, H.A., Masoro, E.J. Nutrition-aging influence on catecholamine-promoted lipolysis. *Metabolism* 29:438–444, 1980.
23. McCarter, R.J.M., Masoro, E.J., Yu, B.P. Rat muscle structure and metabolism in relation to age and food intake. *Am. J. Physiol.* 242:R89–R93, 1982.
24. Herlihy, J., Yu, B.P. Dietary manipulation of age-related decline in vascular smooth muscle. *Am. J. Physiol.* 238:H652–H655, 1980.
25. Weindruch, R.H., Kristie, J.A., Cheney, K.E., Walford, R.L. Influence of controlled dietary restriction on immunologic function and aging. *Fed. Proc.* 38:2007–2016, 1979.
26. Stuchlikova, E., Juricova-Horokova, M., Deyl, Z. New aspects of the dietary effects of life-prolongation in rodents: what is the role of obesity in aging? *Exp. Gerontol.* 10:141–144, 1975.
27. Weindruch, R.H., Kristie, J.A., Walford, R.L. Dietary restriction imposed in middle age mice: life span, disease, immunologic effect. *Gerontologist* 20(5, part II):223, 1980.
28. Sacher, G.A. Life table modification and life prolongation. In Finch, C.E., Hayflick, L. (eds.), *Handbook of Biology of Aging.* New York: Van Nostrand Reinhold, 1977, pp. 582–638.
29. Nohl, H., Hegner, D. Do mitochondria produce oxygen radicals in vivo? *Eur. J. Biochem.* 82:563–567, 1978.

Pathoanatomy of Aging: An Overview

▶

Aubrey J. Hough, Jr., M.D.

4

The separation of pure aging phenomena from intercurrent and accompanying disease states remains very difficult, even with the advances in quantitative biology of the previous quarter century. Although human survivorship has dramatically improved over the last 80 years, the asymptote at around 90 years of age has not changed [1]. Functional measurements of metabolic, respiratory, and musculoskeletal response to stress invariably demonstrate a decline in efficiency in the fourth decade of life [2]. There thus appears to be a finite limitation on human life span independent of intercurrent or age-associated disease [2]. However, a rationale for this remains elusive. The following discussion attempts to organize aging changes, especially those of the skeletal system, into a framework for logical discussion.

No group of criteria for defining aging phenomena has gained universal acceptance. However, requiring the phenomenon in question to be universal, intrinsic, progressive, and deleterious [3] has gained considerable popularity. By adapting these criteria, it is possible to develop a rationale to explain the anatomic lesions that are observed with increased frequency in the aging population [4]. Several lines of reasoning have been employed. One popular approach considers diseases in the aging population as aging-concomitant, aging-dependent, or aging-independent.

AGING-CONCOMITANT DISEASE

Aging-concomitant states are merely accentuations of normal processes that accompany aging in every individual. Examples of such processes include osteoporosis, myofibril atrophy, and neuronal loss. Only when these universal anatomic phenomena attain threshold levels do clinical disease states result. Osteoporosis is regarded by many as an invariable accompaniment to aging [5, 6], with relatively massive losses of bone occurring by the eighth decade [5]. The bone loss in aging is associated with increased resorption while formation proceeds at a relatively normal and constant rate [7]. Sensitivity to parathyroid hormone has been postulated in some types of patients with osteoporosis [7]. More recently, deficient production of 1,25-dihydroxyvitamin D in the elderly [8] has been demonstrated. Whether this phenomenon is directly related to the genesis of osteoporosis is unclear. It is obvious that certain other environmental factors, such as immobilization, calcium-deficient diet, corticosteroids, and castration, can modify and precipitate clinical osteoporosis [5], but the critical definition of universality [3] is maintained, since some degree of osteoporosis is present in every aged individual.

AGING-DEPENDENT DISEASE

Another group of disease states in aging is described as aging-dependent. This group includes several common conditions that, although more common in the aged, are neither universal nor limited to the aged. Examples of this category include chondrocalcinosis [9] and primary osteoarthritis [10]. In other words, only a portion of elderly individuals manifest chondrocalcinosis or osteoarthritis. Their relationship to aging, although statistically valid, may not be a direct one. Thus, age-related changes described in articular cartilage, including loss of chondroitin-4-sulfate [6] and increase in chondrocyte DNA [11], deformability [12], and fatigue fracture [13], may or may not be related to clinical osteoarthritis. Although minor degrees of osteoarthritis are very common in the elderly, clinically significant disease is much less so. Although trauma to joints is associated with the development of osteoarthritis, attempts to show that runners [14] or parachutists [15] develop significantly more osteoarthritis have not been successful, casting doubt that repetitive use produces disease in nonsusceptible individuals. In addition, many aged individuals have perfectly normal joints at autopsy, making any universal aging defect less plausible.

AGING-INDEPENDENT DISEASE

A third group of diseases involves conditions in the aged that result more from cumulative environmental interaction than from intrinsic aging phenomena. Examples of this group include atherosclerosis, secondary osteoarthritis, and osteomalacia. In secondary osteoarthritis, antecedent conditions such as metabolic or inflammatory disease result in severe localized osteoarthritis. Osteomalacia results from a variety of causes, including vitamin D deficiency, renal insufficiency, and impaired absorption of calcium and phosphorus. It is considerably more common in the aged than in the general population [16] but cannot be assumed to be a disease of aging. Numerous other examples of relatively age-independent diseases are apparent, including atherosclerosis, which shows a wide range depending on various dietary, hereditary, and social factors.

CONCLUSIONS

Most diseases associated with aging may not be directly related to aging phenomena but are only indirect manifestations of aging. Comparatively few diseases are universal accompaniments of the aging process. Osteoporosis, myofibril atrophy, and neuronal loss are examples of constants in the aging process. Osteoarthritis, in contrast, is not universal in the aged, only common; thus a direct relationship to aging may not exist. Still other common diseases, such as atherosclerosis, depend more on the quantity of deleterious environmental interactions than on aging, since they affect individuals of a wide age range.

REFERENCES

1. Strehler, B.L. Implications of aging research for society. *Fed. Proc.* 34:5–8, 1975.
2. Goldstein, S. New wrinkles on old age: a cellular and molecular approach. *J. Arkansas Med. Soc.* 78:238–244, 1981.
3. Strehler, B.L. *Time, Cells, and Aging.* New York: Academic Press, 1962.
4. Howell, T.H. *Old Age.* London: H.K. Lewis, 1975, p. 115.
5. Exton-Smith, A.N. The musculoskeletal system. A. Bone aging and metabolic bone disease. In Brocklehurst, J.C. (ed.), *Textbook of Geriatric Medicine and Gerontology.* New York: Churchill Livingstone, 1978.
6. Hall, D.A. *The Aging of Connective Tissue.* New York: Academic Press, 1976, p. 51.
7. Towsey, J., Kelly, P.J., Riggs, B.L., et al. Quantitative microradiographic studies of normal and osteoporotic bone. *J. Bone Joint Surg.* [Am] 47A:785–806, 1965.
8. Slovik, D.M., Adams, J.S., Neer, R.M., et al. Deficient production of 1,25-dihroxyvitamin D in elderly osteoporotic patients. *N. Engl. J. Med.* 305:372–374, 1981.
9. McCarty, D.J. Calcium pyrophosphate crystal deposition disease (pseudogout; articular chondrocalcinosis). In McCarty, D.J. (ed.), *Arthritis and Allied Conditions*, 9th ed. Philadelphia: Lea & Febiger, 1979.
10. Sokoloff, L. *The Biology of Degenerative Joint Disease.* Chicago: University of Chicago Press, 1968.

11. Mankin, J.J., Dorfman, H., Lipiello, L., Zainis, A. Biochemical and metabolic abnormalities from osteoarthritic human hips: 2. Correlation of morphology with biochemical and metabolic data. *J. Bone Joint Surg.* [*Am*] 53A:523–537, 1971.
12. Gardner, D.L. Aging of articular cartilage. In Brocklehurst, J.C. (ed.), *Textbook of Geriatric Medicine and Gerontology*. New York: Churchill Livingstone, 1978, p. 524.
13. Weightman, B. *In vitro* fatigue testing of articular cartilage. *Ann. Rheum. Dis.* 34 (Suppl. 2):108–110, 1975.
14. Puranen, J., Ala-Ketola, L., Peltokallis, P., Saarela, J. Running and primary osteoarthritis of the hip. *Br. Med. J.* 2:424–425, 1975.
15. Murray-Leslie, C.F., Lintoll, D.J., Wright, V. The knee and ankles in sport and veteran military parachutists. *Ann. Rheum. Dis.* 36:327–331, 1977.
16. Sokoloff, L. Occult osteomalacia in American (USA) patients with fracture of the hip. *Am. J. Surg. Pathol.* 2:21–30, 1978.

The Biology of Aging Human Collagen

▶

LeRoy Klein, M.D., Ph.D.
Jess C. Rajan, Ph.D.

5

The aging of collagen develops in parallel with chronologic age, and the physiologic and biomechanical effects of aging have been well established. Some of the questions regarding the role of age-dependent or age-related changes in the connective tissue components of the musculoskeletal system are (a) what is the nature of these changes, (b) are there interorgan differences, and (c) are these changes regarded as normally extended or abnormal maturation?

The effects of aging on collagen have been observed and characterized by several means: morphologically, chemically, metabolically, mechanically, and biologically. The complexity of the aging process in terms of species and tissue differences in the rate of aging [1], collagen fiber organization, fibril dimensions, chemical types of cross-links involved [2], and rate of metabolic turnover necessitates observing the aging of collagen in as many ways as possible. Also, the chemical and metabolic changes that occur during normal growth, development, and maturation must be distinguished from changes that may accompany senescence [3].

There are a number of reasons why collagen as a major component of the musculoskeletal tissues (i.e., bone, tendon, ligament, cartilage) should be susceptible to aging. Collagen and nucleic acids are large, fibrous macromolecules that are old phylogenetically, and both gradually undergo a large number of covalent modifications *after* biosynthesis of the basic chains. The metabolic turnover of collagen is slower than most proteins in growing tissues and very low in nongrowing tissues, thus allowing collagen to remain in situ for

This research was supported by grants AG-00258 and AG-00361 from the National Institutes of Health.

long periods without degradation and replacement with newly synthesized material. Consequently, collagen remains exposed to various internal (metabolic) and external (environmental) agents over long intervals of time. Chemically reactive side groups (aldehydes) are present in collagen that may react inappropriately among themselves or with other compounds and form modified structures with altered physical and metabolic properties.

In the last decade our concept of the molecular structure of collagen has changed from one type of collagen to the existence of several distinct types of collagen. Collagen from bone, tendon, and ligaments is designated as type I collagen and that from cartilage as type II collagen. Skin, scars, and internal organs contain two major types, I and III collagens. Types I, II, and III collagens are termed *interstitial* collagens [4]. The *basement membrane* collagens, type IV, and the *pericellular* collagens, type V, have been isolated from mature and fetal basement membranes. Thus, the different chemical types of collagens reflect the different morphologic types of connective tissues. However, most of the collagen in the body and skeleton is type I collagen.

MORPHOLOGIC OBSERVATIONS

Light Microscopy
Collagen in human bone during embryonic life forms a delicate interwoven network of fibers, which assume a parallel arrangement when lamellar bone develops [5]. With the development of haversian osteons, fiber bundles of collagen are seen to branch out and interconnect with other fiber bundles [6]. As the osteons become more dense and closely packed with age, the fiber bundles undergo a similar rearrangement.

In the noncalcified tissues, such as human skin, the changes of collagen morphology with age are more easily seen. In infants, collagen consists of a loose network of poorly staining young fibers surrounded by numerous cells [7]. Thereafter the collagen fibers become more closely organized and homogeneous in appearance, maintaining a wavy contour, and are relatively acellular. After 60 years of age, collagen fibers are thinned, more elongated, and stretched out so that their wavy appearance is lost.

Electron Microscopy
Collagen fibers of human bone demonstrate variations in size and arrangement throughout life [8]. The overall picture shows that the collagen fibrils increase in diameter and become more closely packed with advancing age. In general, the change in structure and arrangement of the fibrils in one

The Biology of Aging Human Collagen

kind of connective tissue is similar to those of collagenous tissues in other sites of the musculoskeletal system.

The collagen fibrils of infant rib are loosely arranged and have diameters varying from 15 to 60 nm. In mature subjects, the fibrils are more closely packed than in the infant bone and have a larger diameter, which is fairly uniform at 80 to 100 nm. In senile bone, the fibrils are packed more closely, and the diameter is larger and varies between 100 and 150 nm. Similar large fibril diameters that increase with age have been seen in human tendon, 130 to 170 nm [9]; horse and rat tendon, 170 nm [10, 11]; and horse ligament, 150 nm [10]. Smaller fibril diameters are found in young human cartilage, 20 nm [12], and human cornea, 25 nm [13]. With age the human cartilage fibrils increase in diameter to 60 to 100 nm [14], while the human cornea collagen fibrils do not increase in size [15].

The data on fibril diameter (Table 5-1) illustrate that the fibril size, uniformity of size, and effect of age on fibril size varies with each tissue. Thus, aging of collagen may induce different morphologic effects in each tissue at the ultrastructural level.

Table 5-1

Comparison of Collagen Fibril Diameters with Age

| Age | Human Tissues (nm)[a] ||||
	Bone	Tendon	Cartilage	Cornea
Young	15–60	30–40	20–25	25–30
Mature	80–100	100–120	40–50	25–30
Old	100–150	130–170	60–100	25–30

[a]As observed by conventional electron microscopy.

Freeze-Fracture-Etching

The new freeze-fracture-etching technique [16] when applied to collagenous tissues permits a quantitative measurement of the in situ diameters of hydrated collagen fibrils and subfibrils. The ratio of the fibril diameter to the subfibril diameter determines the number of subfibrils per fibril (Table 5-2). With this technique, all collagen fibrils (rat tendon, rabbit cartilage, and cornea) appear closely packed and are two to three times thicker than what has been observed by conventional techniques with the electron microscope.

With age both the diameter of the fibril and the diameter

of the subfibril increase simultaneously, while maintaining their ratio constant. This constant indicates that the thick tendon fibrils (260 nm) contain 100 subfibrils per fibril and the thin cartilage fibrils (48 nm) and corneal fibrils (52 nm) contain only 25 subfibrils per fibril. On the basis of fibril size, its

Table 5-2
Age-Dependence of Collagen Fibril and Subfibril Diameters, and Their Ratios as Revealed by Transverse Freeze-Fracture-Etching Technique

Sample	Age of Animal	Fibril n	Fibril Diameter (nm)[a]	Subfibril n	Subfibril Diameter (nm)[a]	Fibril-Subfibril Diameter Ratio
Rat tail tendon	3 weeks	57	140 ± 40	130	15 ± 6	9.3
	4 months	75	250 ± 60	70	24 ± 6	10.4
	18 months	53	260 ± 40	530	27 ± 10	9.6
Rabbit cornea	3 weeks	71	52 ± 15	128	10 ± 3	5.2
	5 months	85	78 ± 17	78	16 ± 4	4.9
Rabbit cartilage	3 weeks	65	48 ± 15	120	11 ± 3	4.4

[a]Diameters are presented as means ± standard deviation when n equals the number of fibrils or subfibrils counted.

rate of increase in size, and the number of subfibrils per fibril, the effect of aging appears more pronounced on the tissues that contain the thicker fibrils, such as tendon and bone than those tissues that contain the thinner fibrils such as cornea and intestine [16a].

Fibril Dissociation

The exposure of collagenous tissues to dissociative solvents such as 8 M urea or 4 M guanidine hydrochloride followed by electron microscopy has shown a complete dissocation of rat collagen fibrils into subfibrils, filaments, and molecules of collagen [17]. In contrast, a similar study with human dermis [18] has shown varying degrees of dissociation of fibrils into subfibrils depending on the age of the sample. Dermis from an infant, like that seen with rat collagen, showed complete dissociation of all the fibrils but only into subfibrils. Dermal collagen from a 70-year-old man showed only partial dissociation, with some fibrils remaining completely intact and others partially dissociated into subfibrils. A similar observation of intact to partial dissociation of fibrils has been made [19] on adult human intestine after acetic acid extraction. These studies have demonstrated that in long-lived species like humans the collagen interactions (including cross-links) can exist at higher levels of organization than the molecular level and that with age these interactions can in turn increase fibrillar interactions. In lower species (rodents such as mice, rats, or guinea pigs), collagen interactions appear to occur at lower levels of organization so that weaker dissociative solvents such as 0.5 M acetic acid or 0.5 M acidic citrate buffer can solubilize substantial amounts of collagen into filaments and molecules. These weaker dissociative solvents have little effect on collagen from higher species such as human, cow, or dog.

STRUCTURE AND CHEMISTRY

Changes with Age

Collagen in connective tissues is known to undergo several posttranslational changes with age. While the changes that occur during growth and maturation improve the quality of collagen and thus the function of the tissues, the continued progression of these changes after maturation has an adverse effect on the proper functioning of tissues and organs. Formation of excessive cross-links in collagen after maturation is considered to be a major cause of the altered physical, chemical, and metabolic properties of connective tissues during aging.

Reducible Cross-links

The reducible cross-links in collagen originate from lysine and hydroxylysine residues in the molecule. Some of these amino acid residues are metabolically converted (oxidatively deaminated) by the extracellular enzyme lysyl oxidase to compounds known as allysine and hydroxyallysine. Two allysine residues in the N-terminal region of collagen may undergo aldol condensation to form an intramolecular cross-link in young growing tissue. The intermolecular bifunctional cross-links in collagen are formed as Schiff's bases by the reaction of amino and aldehyde groups of allysine and hydroxyallysine [20]. It is not possible to investigate these cross-links directly because of the lability of the aldimine bonds to acid hydrolysis. Therefore, these compounds are reduced with radioactive $NaBH_4$ and hydrolyzed with acid to isolate their stable radioactive derivatives. It has been shown [21] that these reducible cross-links decrease with maturation, and they are virtually absent in aged tissues. To explain this observation, it has been suggested that the reducible cross-links originally present in tissues are stabilized by in vivo reduction or oxidation. The most abundant $NaBH_4$-reducible cross-link in type I collagen of bovine tendon, bone, and dentine is dihydroxylysinonorleucine. However, the molecular distributions of this cross-link in these three tissues are different [22]. Several multifunctional intermolecular cross-links have been found in collagen hydrolysates. These include the trifunctional compounds hydroxymerodesmosine, aldolhistidine, and a tetrafunctional compound, histidinohydroxymerodesmosine (Table 5-3). Their roles in collagen aging are currently unknown.

Table 5-3. Some of the Reducible and Nonreducible Cross-links Found in Collagen

Reducible Cross-links
  Hydroxylysinonorleucine
  Dihydroxylysinonorleucine
  Hydroxymerodesmosine
  Aldolhistidine
  Histidinohydroxymerodesmosine

Nonreducible Cross-links
  Pyridinoline
  Deoxypyridinoline
  Lysinoalanine
  Histidinoalanine

Nonreducible Cross-links

Recently Fujimoto [23] reported the isolation of a fluorescent cross-linking compound from bovine Achilles tendon and bone collagens without prior reduction. This trifunctional compound has been given the trivial name pyridinoline, because it is a 3-hydroxypyridinium derivative with three amino and three carboxyl groups. Pyridinoline is considered to be derived by the condensation of hydroxylysinohydroxynorleucine. Its synthesis appears to be dependent on lysyl oxidase and occurs at a slower rate as compared with the synthesis of bifunctional reducible cross-links [24]. Pyridinoline content in rat tissues has been found to continue to increase after maturity, but a decrease has been noted in human tissues after about 30 years of age. The detection of relatively high concentrations of pyridinoline in human cartilage and bone has led to the suggestion that it is a mature cross-link in hard tissue collagens and thus may play a role in aging.

New Class of Nonreducible Cross-links

Two new nonreducible cross-links, lysinoalanine and histidinoalanine, have been reported [25] to be present in hard tissue collagens. Because their synthesis does not require lysyl oxidase, they represent a new class of nonreducible cross-linking compounds. The content of histidinoalanine was found to increase progressively with age in human connective tissues [25].

METABOLISM

Biosynthesis

Intracellular biosynthesis of collagen [26] involves the formation of a relatively large precursor molecule known as procollagen. The procollagen chains contain noncollagenous sequences at the amino and carboxy terminal ends. The collagen peptide chains after their synthesis undergo several intracellular modifications. These include hydroxylation of certain prolyl and lysyl residues, glycosylation of some hydroxylysine residues, and interchain disulfide bond formation. Prolyl hydroxylation is an essential step in the formation of the collagen triple helix. These intracellular modifications are mediated by the enzymes prolyl hydroxylase, lysyl hydroxylase, galactosyl transferase, and glucosyl transferase.

Conversion of procollagen to collagen occurs extracellularly. For types I, II, and III collagens, the propeptides are removed by two specific procollagen peptidases, one for each end. Each collagen type is believed to have its own specific peptidases. Following this conversion of procollagen to collagen, the protein is deposited in fibrous form extracellularly. The primary structure of collagen is then further modified by a copper-containing enzyme, lysyl oxidase, which converts certain lysine and hydroxylysine side chains to aldehydes. These aldehyde groups participate in the formation of intra- and intermolecular cross-links through a series of spontaneous condensation reactions. None of these early biosynthetic steps appears to be directly involved in the aging of collagen.

Metabolic Turnover

While the structural aspects of connective tissues have made significant contribution to the understanding of aging, little input has come from the kinetic aspect. This is mainly due to the absence of sensitive and quantitative methods for measuring collagen turnover in slowly metabolizing pathways. Most of our current knowledge on collagen turnover has been derived from relatively active metabolic conditions like normal growth, wound healing, or disease. This turnover in radioactive animals can be quantified in terms of *old* radioactive ^{3}H-collagen loss and *new* nonradioactive ^{1}H-collagen replacement [27] and their relationship to each other with aging.

Turnover rates for collagen vary, often widely, from tissue to tissue during growth [28, 29] but become almost static during maturation. In adult rats following denervation of the hind limb, there is a small but significant turnover of collagen in tendons, ligaments, menisci, and bones [30]. This is in sharp contrast to collagen's metabolic inertness in normal adult rats. However, the turnover due to paralytic disuse is less than that observed in rapidly growing rats [28]. Increased collagen turnover in adult tissues would mean that significant amounts of newly synthesized collagen with fewer cross-links would be present, and this would reduce the mechanical strength of the tissues. It is of interest that the newly synthesized collagen in the adult limb after denervation or fracture healing contains only type I collagen in bone and a mixture of types I and III collagens in scar tissue.

MECHANICAL PROPERTIES
Bone Collagen

The contribution of collagen to the mechanical properties of bone can be understood in terms of a three-dimensional arrangement of collagen bundles [6] and fibrils and their rearrangement with age. In addition, changes in the relative concentrations of bone collagen and mineral with age must be considered.

Scanning electron-microscopic pictures of all bone matrix surfaces have shown that collagen fiber bundles are not discrete but branch [31]. Collagen fibrils in one bundle branch off into other bundles laterally and vertically, so that the collagen bundles act as a continuum. The strength of the collagen matrix as a continuum will depend to a large extent on the strength of the lateral bonding between fibrils and between subfibrils. Also, the strength of the bone matrix will depend on the tensile strength of the collagen bundles and on the ground substances that bind fiber bundles into larger composites. Thus, the ultimate strength of collagen is dependent on the sum of interactions at different levels of organization.

The arrangement of collagen fibers in lamellae of human osteons changes with age [6]. This change consists of an increase in the number and concentration of longitudinal collagen fibers, until they are no longer confined to exact lamellae, and of a simultaneous decrease in the number and concentration of circumferential collagen fibers. Thus, with age, the number of circumferential lamellae is reduced, and the number of longitudinal lamellae per osteon is increased. In the areas of bone having the greatest resistance to fracture (and the highest modulus of elasticity), the majority of osteons have longitudinal or steeply spiraling collagen fibers. The strongest specimens of bone, in regard to the tensile breaking load, show that the majority of the collagen fibers are oriented longitudinally. Counteracting the improved three-dimensional arrangement of collagen fibers with age are the observations that the osteons are shorter and narrower, while the haversian canals are wider. This leads to a decrease in size of haversian systems with age and an increase in the proportion of extrahaversian bone [32].

Bone Collagen and Mineral
The contribution of collagen and mineral was examined by studying the mechanical properties of bone under tensile loading after partial decalcification [33]. In determining the effect of age on the elastic-plastic properties of bone [33], it appeared that the slope of the elastic region was governed mostly by the mineral component and the slope of the plastic (nonelastic) region was governed by only the collagen component. The age-related increase in the elastic slope that was observed was consistent with increasing deposition of the mineral phase. The slope of the nonelastic response (plastic zone), being characteristic of the stiffness of the collagen component, would be expected to increase progressively as collagen ages and adds more cross-links. For a human population ranging in age from 21 to 89 years, an increase was observed in the plastic slope of femoral bone [34], indicating increased stiffness in the plastic zone as a reflection of the aging effect on collagen structure.

The most important age change that occurs in the tensile properties of bone is the decrease in the plastic properties and hence in the plastic modulus, total or ultimate strain before failure, and energy absorption [34]. The decrease in energy absorption to impact with age appears to be caused partially by an increased mineralization of bone [35]. The increased porosity in old bone appears to exacerbate the effect of high mineralization. The high mineral content reduces the ability of bone to undergo plastic deformation with age.

BIOLOGIC REACTIVITY

Response to Trauma and Repair
Experimentally produced fractures or skin wounds are known to heal more quickly in young animals than in old ones. Formation of fibrous tissue in fracture callus [36] is slower in old than in young mice. However, qualitatively human scar tissues appear to have their own biologic age [37], which is independent of the chronologic age of the individual, although a certain correlation exists between the two. In general, older individuals form larger quantities of collagenous materials than the young but at a slower rate.

Therefore, it is essential to distinguish between the rate of the process and the quantitative aspects of mass [38]. Old humans [39] as well as old rats produce more collagen in the granulation tissue of a polyvinyl sponge implant than do their younger counterparts. These studies measure the animal's response to the continual presence of a foreign body and may involve factors other than those operative in the repair of an incisional skin wound or fracture callus.

In rats, repair of a skin wound in a younger rat produces more collagenous scar tissue than in an old animal. In addition to quantitative differences in the amount of fibrous tissue deposited in a skin wound, there may be qualitative differences in the mechanical behavior of the induced tissue [40]. The regenerated tissue in the older animal may become mechanically stiffer than that formed in the younger animal, although early in the wound healing period the regenerating tissues from young and old animals have a similar mechanical response to applied stress.

CONCLUSIONS

The understanding of the aging of human collagen is still in its infancy. Many of the collagen cross-links studied heretofore appear to be derived from young tissues of rodents and may not play any major role in the aging of human tissues. Newly discovered cross-links in human tissues, such as pyridinoline in cartilage, histidinoalanine and lysinoalanine in calcified tissues, may be involved in aging of hard tissues. The major ultrastructural site of cross-links needs to be resolved in terms of being intrafibrillar, interfibrillar, or interfiber, or perhaps all three sites are involved. Since so many modifications of collagen structure can occur extracellularly, it is not clear whether cells are directly involved in the aging process or whether cells indirectly play a role through the biosynthesis of appropriate extracellular enzymes.

The aging of bone cannot be explained by a single mechanism. It appears to involve the additional cross-linking of collagen, the architectural rearrangement of collagen fibers and osteons, and the hypermineralization of bone matrix.

REFERENCES

1. Hamlin, C.R., Luschin, J.H., Kohn, R.R. Aging of collagen: comparative rates in four mammalian species. *Exp. Gerontol.* 15:393–398, 1980.
2. Bailey, A.J. Tissue and species specificity in the crosslinking of collagen. *Pathol. Biol.* (Paris) 22:676–680, 1974.
3. Bornstein, P., Traub, W. The chemistry and biology of collagen. In Neurath, H., Bailey, K. (eds.), *The Proteins*, vol. IV. New York: Academic, 1979, pp. 411–632.

4. Viidik, A. Age-related changes in connective tissues. In Viidik, A. (ed.), *Lectures on Gerontology*, vol. 1A. London: Academic, 1982, pp. 173–211.
5. Silberberg, M., Silberberg, R. Ageing changes in cartilage and bone. In Bourne, G.H. (ed.), *Structural Aspects of Ageing*. London: Pitman, 1961, pp. 85–108.
6. Smith, J.W. The arrangement of collagen fibers in human secondary osteons. *J. Bone Joint Surg. [Br]* 42B:588–605, 1960.
7. Hill, W.R., Montgomery, H. Regional changes and changes caused by age in the normal skin. *J. Invest. Dermatol.* 3:231–245, 1940.
8. Robinson, R.A., Watson, M.I. Crystal-collagen relationships in bone as observed in the electron microscope: III. Crystal and collagen morphology as a function of age. *Ann. N.Y. Acad. Sci.* 60:596–628, 1955.
9. Pahlke, G. Electron microscopic study of the intercellular substance of human tendon. *Z. Zellforsch.* 39:421–430, 1954.
10. Parry, D.A.D., Craig, A.S., Barnes, G.R.G. Tendon and ligament from the horse: an ultrastructural study of collagen fibrils and elastic fibers as a function of age. *Proc. R. Soc. Lond. [Biol.]* 203:293–303, 1978.
11. Torp, S., Baer, E., Friedman, B. Effects of age and mechanical deformation on the ultrastructure of tendon. In Atkins, E.D.T., Keller, A. (eds.), *Structure of Fibrous Biopolymers*. London: Butterworth, 1975, pp. 223–250.
12. Bonucci, E. The locus of initial calcification in cartilage and bone. *Clin. Orthop.* 78:108–139, 1971.
13. Schwarz, W., Graf Keyserlingk, D. Electron microscopy of normal and opaque human cornea. In Langham, M. (ed.), *The Cornea*. London: Johns Hopkins Press, 1969, pp. 123–131.
14. Minns, R.J., Steven, F.S. The collagen fibril organization in human articular cartilage. *J. Anat.* 123:437–457, 1977.
15. Kanai, A., Kaufman, H.E. Electron microscopic studies of corneal stroma: aging changes of collagen fibers. *Ann. Ophthalmol.* 5:285–292, 1973.
16. Itoh, T., Klein, L., Geil, P.H. Age dependence of collagen fibril and subfibril diameters revealed by transverse freeze-fracture and -etching technique. *J. Microsc.* 125:343–357, 1982.
16a. Klein, L., Eichelberger, H., Mirian, M., Hiltner, A. Ultrastructural properties of collagen fibrils in rat intestine. *Conn. Tissue Res.* 12:71–78, 1983.
17. Ruggeri, A., Benazzo, F., Reale, E. Collagen fibrils with straight and helicoidal microfibrils: a freeze-fracture and thin-section study. *J. Ultrastruct. Res.* 68:101–108, 1979.
18. Lillie, J.H., MacCallum, D.K., Scalatta, L.J., Occhino, J.C. Collagen structure: evidence for a helical organization of the collagen fibril. *J. Ultrastruct. Res.* 58:134–143, 1977.
19. Steven, F.S., Jackson, D.S., Schoefield, J.D., Bard, J.B.L. Polymeric collagen isolated from the human intestinal submucosa. *Gut* 10:484–487, 1969.
20. Bailey, A.J., Robins, S.P., Balian, G. Biological significance of the intermolecular crosslinks of collagen. *Nature* 251:105–109, 1974.
21. Robins, S.P., Shimokomaki, M., Bailey, A.J. The chemistry of collagen crosslinks: age-related changes in the reducible components of intact bovine collagen fibers. *Biochem. J.* 131:771–780, 1973.
22. Kuboki, Y., Mechanic, G.L. Comparative molecular distribution of crosslinks in bone and dentine collagen: structure-function relationships. *Calcif. Tissue Int.* 34:306–308, 1982.
23. Fujimoto, D. Isolation and characterization of a fluorescent material in

bovine Achilles tendon collagen. *Biochem. Biophys. Res. Commun.* 76:1124–1129, 1977.
24. Siegel, R.C., Fu, J.C.C., Uto, N., et al. Collagen crosslinking: lysyl oxidase dependent synthesis of pyridinoline in vitro: confirmation that pyridinoline is derived from collagen. *Biochem. Biophys. Res. Commun.* 108:1546–1550, 1982.
25. Fujimoto, D., Hirama, M., Iwashita, T. Histidinoalanine, a new crosslinking amino acid in calcified tissue collagen. *Biochem. Biophys. Res. Commun.* 104:1102–1106, 1982.
26. Eyre, D.R. Collagen: molecular diversity in the body's protein scaffold. *Science* 207:1315–1322, 1980.
27. Klein, L., Lewis, J.A. Simultaneous quantification of ^{3}H-collagen loss and ^{1}H-collagen replacement during healing of rat tendon grafts. *J. Bone Joint Surg.* [*Am*] 54A:137–146, 1972.
28. Klein, L., Zika, J.M. Comparison of whole calvarial bones and long bones during early growth in rats: II. Turnover of calcified and uncalcified collagen masses. *Calcif. Tissue Res.* 20:217–227, 1976.
29. Klein, L., Rajan, J.C. Collagen degradation in rat skin but not in intestine during rapid growth: effect on collagen types I and III from skin. *Proc. Natl. Acad. Sci. USA* 74:1436–1439, 1977.
30. Klein, L., Dawson, M.H., Heiple, K.G. Turnover of collagen in the adult rat after denervation. *J. Bone Joint Surg.* [*Am*] 59A:1065–1067, 1977.
31. Boyde, A. Scanning electron microscope studies of bone. In Bourne, G.H. (ed.), *The Biochemistry and Physiology of Bone*, vol. 1. New York: Academic, 1972, pp. 259–310.
32. Currey, J.D. Some effects of ageing in human Haversian systems. *J. Anat.* 98:69–75, 1964.
33. Burstein, A.H., Zika, J.M., Heiple, K.G., Klein, L. Contribution of collagen and mineral to the elastic-plastic properties of bone. *J. Bone Joint Surg.* [*Am*] 57A:956–961, 1975.
34. Burstein, A.H., Reilly, D.T., Martens, M. Aging of bone tissue: mechanical properties. *J. Bone Joint Surg.* [*Am*] 58A:82–86, 1976.
35. Currey, J.D. Changes in the impact energy absorption of bone with age. *J. Biomech.* 12:459–469, 1979.
36. Tonna, E.A. Fracture callus formation in young and old mice observed with polarized light microscopy. *Anat. Rec.* 150:349–362, 1964.
37. Verzar, F., Willenegger, H. Aging of the collagen in skin and scars. *Schweiz. Med. Wochenschr.* 41:1234–1236, 1961.
38. Chvapil, M. Regulation of the growth of connective tissue in organs and tissues: skin. In *Physiology of Connective Tissue*. London: Butterworth, 1967, pp. 248–249.
39. Boucek, R.J., Noble, N., Kao, K.Y.T., Elden, H.R. The effects of age, sex, and race upon the acetic acid fractions of collagen. *J. Gerontol.* 13:2–9, 1958.
40. Sussman, M.D. Aging of connective tissue: physical properties of healing wounds in young and old rats. *Am. J. Physiol.* 224:1167–1171, 1973.

The Biology of Aging Muscle: Quantitative Versus Qualitative Findings of Performance Capacity and Age

▶

Peter Jokl, M.D.

All the world's a stage, and all the men and women merely players. They have their exits and their entrances, and one man in his time plays many parts, his acts being seven ages.
—Shakespeare, *As You Like It*

This chapter is divided into two sections. The first section is a summary of measured data describing changes in muscle associated with the aging process. In contrast, the second part presents information from performance measurements in gerontologic populations as seen in sports medicine and epidemiologic occupational performance reports.

On the quantitative level, one can state that all measurements of muscle, be they physiologic, anatomic, histochemical, or enzymatic, decrease after age 40. Many of the aging phenomena indicate a decrease in the number of active functional units (i.e., motor units or muscle fibers) and a loss in concentration of specific enzymes or fiber types. These studies represent quantitative data and do not take into consideration the effect of these changes on qualitative performance levels, which are well exemplified in practical clinical sports medicine. Nature, it appears, has constructed its normal biologic units with adequate reserve to maintain excellent function despite described losses of physiologic subunits.

It is estimated that subconcussional blows to the central nervous system induce the degeneration of thousands of neural elements, which only with repetition and time produces measurable change in performance [1]. Neurosurgeons at times are forced to remove substantial amounts of neural tissue because of trauma or neoplasm, often with minimal loss or rapid subsequent recovery of intellectual and motor function. Premiere performances by athletes have been accomplished with massive loss or degeneration of skeletal muscle tissue thought to be essential for a maximum effort.

One example is that of a world-record performance in the hammer throw by an athlete who at birth had suffered an injury of both the upper and lower brachial plexuses of the left arm, which caused a marked growth defect as well as generalized paresis and paralysis of the muscles of the extremity [2]. An example of both neural and muscular accommodation is in pistol shooting in a right-handed individual who in the 1936 Olympics placed in the top ten competitors in his class. Shortly thereafter, in a tragic accident, he lost his right upper extremity. He was subsequently able to retrain the right nondominant motor cortex and the left nondominant upper extremity to the degree that he won gold medals in the successive 1948 and 1952 Olympic Games. These cases indicate that biologic systems, and more specifically the neuromuscular system, appear to have tremendous reserves that will allow continued high-level performance despite the loss of a significant number of functional units [2]. This phenomenon must be taken into consideration in interpreting the quantitative data presented as a pattern of the aging biologic system. The early onset and continued use of intellectual and motor function [3] appear to be more important criteria in influencing the function of the neuromuscular system in advanced age. These factors appear to be most crucial in preventing atrophy of function with age.

QUANTITATIVE ASPECTS OF MUSCLE AGING

Physiologic

The quantitative measurements of changes in muscle function with age indicate a general loss of lean body mass, primarily skeletal muscle [4]. Additionally, there is a decline in almost all measurable physiologic functions (Fig. 6-1). It is estimated that 20 to 40 percent of maximal muscle strength is lost by age 65 [5] (Fig. 6-2).

Specific measurements of muscle isometric and isokinetic performances all showed deterioration with age [6, 7]. Interestingly, no change in endurance capacity was correlated with age. Prolongation of reflex half relaxation time occurs with aging [8]. Electromyographic patterns indicate an increasing recruitment of motor units for a given task by skeletal muscle with age. This effect is felt to be due to a general-

The Biology of Aging Muscle

ized denervation of muscle fibers requiring a larger number of motor units to produce a given force [9]. Table 6-1 presents a general summary of physiologic changes in skeletal muscle that are associated with age.

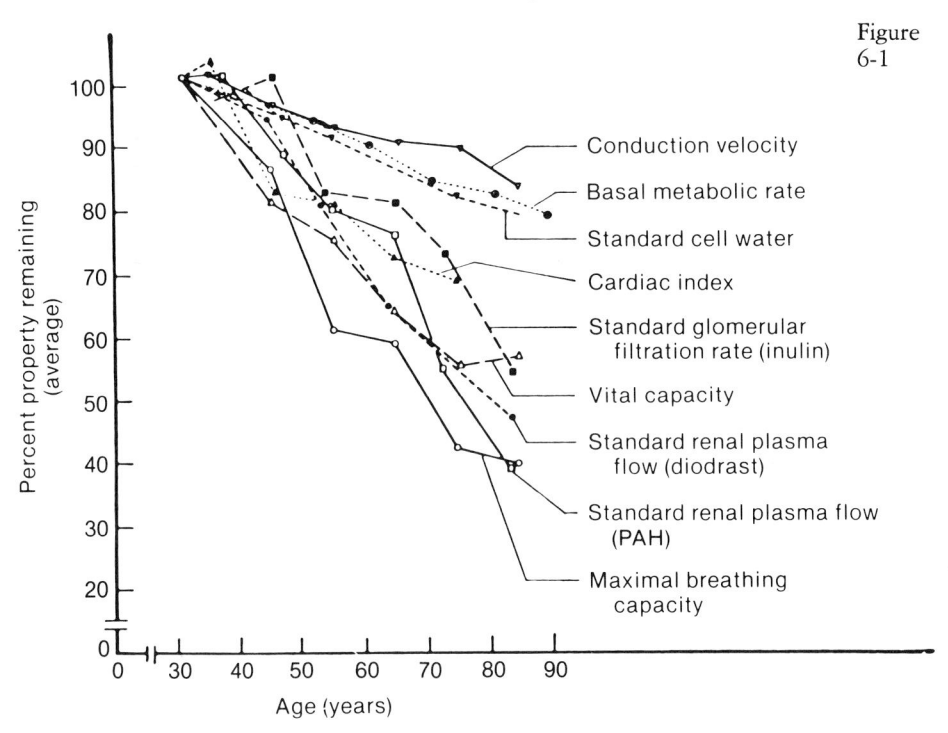

Figure 6-1

Decline in various human functional capacities and physiologic measurements. Values are adjusted so that the value at age 30 equals 100 percent. Other values are expressed as the average percentage of the value at age 30 remaining at the specified age.
SOURCE: Reproduced with permission from Shock, N.W. The science of gerontology. In Jeffers, E.C. (ed.), *Proceedings of Seminars 1959–61, Durham, N.C., Council on Gerontology.* Durham, N.C.: Duke University Press, 1962.

Physiologic Changes with Aging

Table 6-1

Decreased handgrip strength
Decreased functional motor units
Decreased number of muscle fibers
Prolongation of half relaxation time (TA)
Decreased quadriceps strength
Increased electromyographic recruitment
Decreased isometric strength
Increased endurance with age
Decreased speed of muscle contraction
Increased rate of Ca^{++} transport by sarcoplasmic reticulum

SOURCE: Data from Shephard [4], Larsson and Karlsson [7], Carel et al. [8], and Hayward [9].

Anatomic and Histopathologic Changes

Anatomic alterations in skeletal muscle with age present a gross pattern of skeletal muscle wasting. On the microscopic level, pathologic changes similar to those described in myopathy, disuse, and neuropathic atrophy have been noted

Figure 6-2

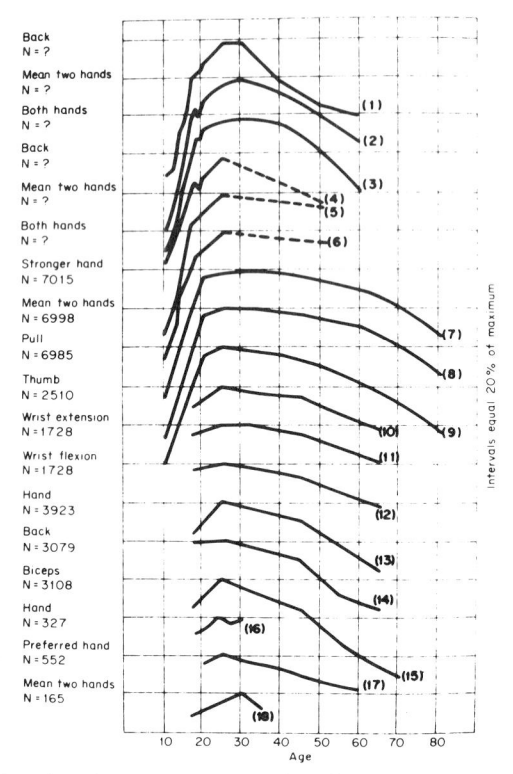

Summary of studies from various periods indicating a general loss of muscle strength after age 40.
SOURCE: Reproduced with permission from Kohn, R.R., *Principles of Mammalian Aging*, 2nd ed. Englewood Cliffs, N.J.: Prentice-Hall, 1978, p. 170.

[10, 11]. These include target fibers, central nuclei, phagocytosis, interstitial fibrosis, nemaline rods, cytoplasmic bodies, and cell infiltrates (Table 6-2).

Ultrastructural examination shows streaming of the Z band, dilatation and increases in the size of the sarcoplasmic reticulum, and widening of cellular and surrounding capillary basement membranes. Patchy myofibrillar degeneration, abnormal satellite cells, increased number of mitochondria, accumulation of lipopigment, and lamellar figures are noted. At the neuromuscular junction, an increase in the number of synaptic vesicles is observed.

The hypertrophy of the sarcoplasmic reticulum is felt to correlate with an increase in the rate of calcium ion transport or reabsorption by the sarcoplasmic reticulum. This effect may in turn cause contractile protein dysfunction due to the paucity of calcium ions available for normal contraction [12] (Table 6-3).

The thickening of the muscle fibers' basement membrane is hypothesized as explaining the decreased depolarization sensitivity of aged muscle cell membranes [12].

Table 6-2

Anatomic Changes with Aging

Loss of lean tissue
Skeletal muscle wasting
Sarcoplasmal nuclei
Central nuclei
Interstitial fibrosis
Ragged red changes
Cell infiltration
Target formation

SOURCE: Data from Tomonaga [10] and Scelsi et al. [11].

Table 6-3

Ultrastructural Changes with Aging

Z-band streaming
Nemaline rod formation
Lamellar figures
Myofilament whorls
Accumulation of lipopigment
Myelin figures
Curvilinear bodies
Deformity of muscle nuclei
Abnormal satellite cells
Increased mitochondria
Dilatation of sarcotubular system
Thickening of capillary basement membrane
Thickened subsynaptic fold
Decreased diameter of type II fibers
Hypertrophy of the sarcoplasmic reticulum

SOURCE: Data from Tomonaga [10], Scelsi et al. [11], and Frolkis et al. [12].

Histochemical studies of aging skeletal muscle are dominated by the finding of a general increase in type I muscle fibers occurring in all skeletal muscles studies [10, 11]. The etiology of this finding is felt to be twofold and represents either a selective loss of type II muscle fibers or a reinnervation of denervated type II fiber motor units by type I motor fibers [10, 11].

Measurement of myosin adenosine triphosphatase (AT-Pase) activity shows a progressive decrease in activity levels with age [14, 15]. Finally, a loss of elastic tissue was noted on selective histochemical studies in aged skeletal muscle (Table 6-4).

Table 6-4

Histochemical Changes with Aging

Decreased myofibrillar ATPase
Increased atrophy of type II fibers
Type I fiber predominance
Conversion to type I fiber
Decreased number of type II fibers
Lower enzyme activity

SOURCE: Data from Larsson and Karlsson [7], Tomonaga [10], Scelsi et al. [11], and Mohler [16].

QUALITATIVE PHYSICAL PERFORMANCE CAPABILITY IN THE HEALTHY ELDERLY POPULATION

Work and sport performance data indicate that a selective sampling of healthy elderly subjects produces remarkable examples of high levels of neuromuscular performance (Fig. 6-3.) Mohler [16], in a paper reviewing Federal Aviation

Figure 6-3

Larry Lewis, age 101, broke a world's record by finishing the 100-yard dash in 17.8 seconds. The record he broke is for athletes 100 years old or over.

The Biology of Aging Muscle

Agency data on pilot performance as correlated with age, found that the 55-year-old and above group carried the best performance records when compared statistically with all other younger aged categories. Of importance to our topic is the selectivity of subjects under study because of the stringent health and performance tests that pilots are required to undergo annually to maintain their certification. As Mohler clearly illustrates, superior performance levels can be expected from "older" subjects as long as disease states, which erroneously have been felt to be synonymous with aging, are eliminated (Fig. 6-4).

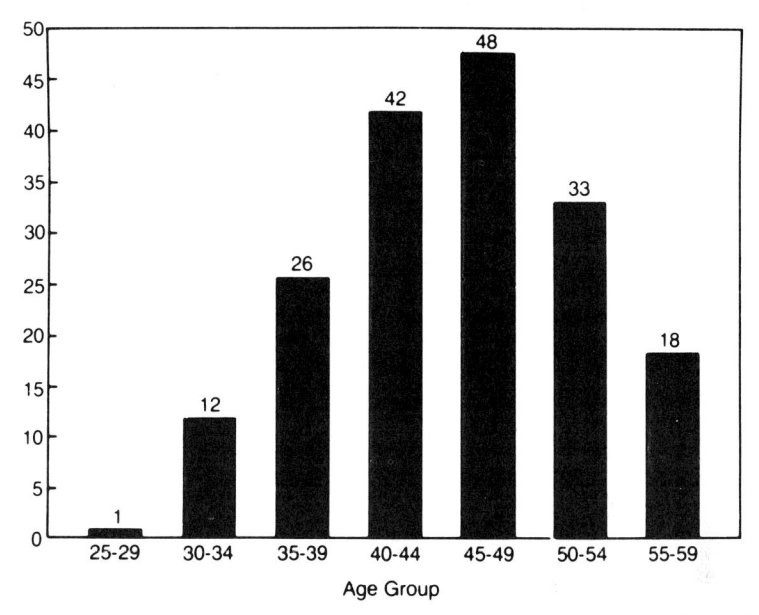

Figure 6-4

Captain age plotted against airline accidents demonstrates a marked falloff after the 40s. According to data from the National Transportation Safety Board, older pilots are more experienced and less likely to be involved in accidents (NTSB Reports). Data are for scheduled domestic passenger service on U.S. certified route air carriers, 1970–1977.
SOURCE: Reproduced with permission from Mohler [16].

Kavanagh and Shephard [4], in the evaluation of group performances in an International Master's Track and Field Championship, found in the 50- to 60- and 60- to 70-year age groups that levels of performance measuring 83 and 81 percent, respectively, of the world record standard in the long-distance running events were achieved in competition. To translate these figures into practical performance standards would indicate a marathon run of 2 hours and 45 minutes for a 70-year-old man, an accomplishment that is untenable for a majority of trained younger individuals and impossible for untrained individuals (Table 6-5).

The accomplishments of complex required routines for participants up to age 90 in the annual West German Master's Gymnastic Competitions are indicated in Figures 6-5 and 6-6. These are examples of performances at the highest levels of skill attainable at an advanced age [17].

Table 6-5. Performance of Male Contestants Relative to Current World Record Speed for Their Age and Distance

	Mean Percent ± S.D.		
Age (yr)	Sprint	Middle Distance	Long Distance
<40	88.8 ± 5.1	—	86.1 ± 10.3
40–50	86.4 ± 5.7	88.3 ± 7.2	82.8 ± 8.1
50–60	88.1 ± 4.6	87.3 ± 6.6	82.9 ± 10.0
60–70	76.6 ± 3.9	73.1 ± 7.1	81.1 ± 6.7
70–80	63.8 ± 3.1	—	—

SOURCE: Kavanagh and Shephard [4], p. 658.

Figure 6-5

Obligatory exercises on the side horse for men aged 50–59 (*top*), on parallel bars for men aged 60–84 (*middle*), and on horizontal bar for women aged 32–52 (*bottom*), performed by several hundred trained participants in the National Gymnastic Festival at Marburg in 1952. The degree of neuromotor differentiation required from the gymnasts was such that healthy young men and women between the ages of 18 and 22 years were unable to perform them unless they had practiced regularly for prolonged periods.
SOURCE: Reproduced with permission from Jokl, E. [2], p. 32.

CONCLUSIONS

The examples of mental and physical achievement in elderly individuals contradict the findings summarized in the quantitative review of data of aging in the first part of this chapter. It is possible that the gerontologic literature is based on studies of subjects selected from hospitals, nursing homes, and other institutional settings that do not represent healthy examples of older individuals [3]. These factors must be considered in order to explain the functional and quantitative discrepancies. Clearly from the findings of clinical and practical sports medicine, normal function can be expected despite the described pathologic, physiologic, anatomic, and histochemical changes noted with age. It is important to distinguish the healthy elderly from those individuals who suffer from specific disease entities. Clement [3] indicates that a determining factor in intellectual and physical performance levels with age appears to be correlated with the early onset in life of physical and intellectual activity and their continued application. These appear to determine functional capacities later in life. Physiologic adaptation to training with improvement in performance levels can occur at any age [4,

Figure 6-6

Senior gymnast performing the required obligatory exercises in the National German Gymnastic Festival. These athletes are examples of maintaining a high degree of athletic competence associated with uninterrupted years of training in gymnastics.

18]. It is apparent that the human biologic system has a substantial reserve capacity that can overcome measured degenerative changes that occur with aging.

REFERENCES

1. Unterharnscheidt, F.J. Injuries due to boxing and other sports. In Vinken, P.J. (ed.), *Handbook of Clinical Neurology,* vol. 23. New York: North Holland Publishing Co., 1975, pp. 527–593.
2. Jokl, E. *Physiology of Exercise.* Springfield, Ill.: Charles C Thomas, 1971, pp. 108–112.
3. Clement, F.J. Longitudinal and cross sectional assessments of age changes in physical strength as related to sex, social class and mental ability. *J. Gerontol.* 29:423–429, 1974.
4. Kavanagh, T., Shephard, R.J. The effects of continued training on the aging process. *Ann. N.Y. Acad. Sci.* 301:656–670, 1977.
5. Shock, N.W. Physical activity and the "rate of ageing." *Can. Med. Assoc. J.* 96:836–840, 1967.
6. Larsson, L., Grimby, G., Karlsson, J. Muscle strength and speed of movement in relation to age and muscle morphology. *J. Appl. Physiol.* 46:451–456, 1979.
7. Larsson, L., Karlsson, J. Isometric and dynamic endurance as a function of age and skeletal muscle characteristics. *Acta Physiol. Scand.* 104:126–136, 1978.
8. Carel, R., Korczyn, A.D., Hochberg, Y. Age and sex dependency of Achilles tendon reflex. *Am. J. Med. Sci.* 279:57–63, 1979.
9. Hayward, M. Automatic analysis of the electromyogram in healthy subjects of different ages. *J. Neurol. Sci.* 23:397–413, 1977.
10. Tomonaga, M. Histochemical and ultrastructural changes in senile human skeletal muscle. *J. Am. Geriatr. Soc.* 25:125–131, 1977.
11. Scelsi, R., Marchetti, C., Poggi, P. Histochemical and structural aspects of m. vastus lateralis in sedentary people (age 65–89). *Acta Neuropathol (Berl.)* 51:99–105, 1980.
12. Frolkis, V.V., Martynenko, O.A., Zamostyan, V.P. Aging of the neuromuscular apparatus. *Gerontology* 22:244–279, 1976.
13. McCarter, R. Effects of age on contraction of mammalian skeletal muscle. In Kaldor, G., Battista, W.J.D. (eds.), *Aging,* vol. 6. New York: Raven Press, 1978, pp. 1–21.
14. Gutmann, E., Hanzlikova, V. Basic mechanisms of aging in the neuromuscular system. *Mech. Ageing Dev.* 1:327–349, 1972.
15. Jennekens, F.G.I., Tomlinson, B.E., Walton, J.N. Histochemical aspects of fiber limb muscles in old age. *J. Neurol. Sci.* 14:259–276, 1971.
16. Mohler, S.R. Reasons for eliminating the "age 60" regulation for airline pilots. *Aviat. Space Environ. Med.* 52:445–454, 1981.
17. Jokl, E. *Alter and Leistung.* Berlin: Springer Publishing, 1954.
18. Barry, A.J., Steinmetz, J.R., Page, H.F., Rodahl, K. The effects of physical conditioning on older individuals: II. Motor performance and cognitive function. *South. J. Gerontol.* 21:192–199, 1966.

The Aging of Articular Cartilage

▶

O. Donald Chrisman, M.D.

This chapter will discuss two questions: (1) what are the effects of age on articular cartilage, and (2) is there a connection between aging and osteoarthritis? The standard, classic answers are, there are remarkably few aging effects, and they are benign, showing little if any relationship to osteoarthritis.

Articular cartilage is a wonderful tissue that can provide low friction surfaces to moving joints for around a hundred years. This kind of self-repairing, self-lubricating performance is a good deal better than General Motors or other industrial concerns are able to produce. The effects of aging in cartilage may produce very little change in function or gross morphology. However, there are rather subtle changes, and these may be more significant than they initially appear.

Let us look at each cartilage component and see what happens with aging.

CELLS

Cell counts show a decrease in all layers of articular cartilage with age. It is rare to see mitoses in adult articular cartilage—so rare that some histologists doubted that they could occur. However, we were able to show mitoses in adult canine cartilage after pressure clamp damage [1]. Johnson [2] makes a good case for mitotic activity in slow joint remodeling, and it would seem to occur in acromegaly. Certainly organelles for mitosis can be seen with the electron microscope in adult cartilage. Granted, however, that there is very little cellular replication in aging cartilage, we do not find that age slows down its metabolic activity. Cell for cell, aging cartilage metabolizes nutrients and remodels its matrix just as rapidly as young cartilage. A very interesting set of experiments by

Roughley and colleagues [3] compares human tibial plateau cartilage (ages 62–81 years) removed from the exposed central area and the more protected zone below the meniscus. With age, the superficial open zone falls to two or three cell layers thick, while the submeniscal area remains at six to eight layers. There is a very obvious surface degeneration in the cartilage directly exposed to weight-bearing stresses as compared to the protected area. However, there is very little change in the cells or matrix below this area of degeneration.

COLLAGEN

Although there is obvious disruption of the superficial collagen layer histologically, there is very little biochemical change. The suggestion has been made that loss of cross-linkages between collagen bundles may be an early change in degeneration and even that an enzyme to be called collagen "linkase" might be present and conceivable as a cause of early damage. No hard data in this direction have been presented.

WATER CONTENT

Water is the major component of cartilage and is present in several pools. It can be bonded chemically, more loosely bound to large molecular matrix components by van der Waal's forces, or relatively free in the intercellular area. With age, water content tends to diminish as cells and other matrix components diminish. With increasing histologic degeneration of cartilage, however, water content increases, leading to the hypothesis that fibrillation, with disruption of the tight collagen network in cartilage, allows swelling of the matrix, explaining the increase in water.

PROTEOGLYCANS

It has been known since Von Kuhn's work [4] in 1958 that there is a gradual shift from a predominance of chondroitin sulfate A in childhood to the combination of chondroitin sulfate C and keratosulfate by age 35 or 40. Santer et al. [3] provide a modern update on proteoglycan changes. Bollet [5] found a decrease in the average length of the carbohydrate chains with age and tried to link this finding with hyaluronidase activity but was never able to demonstrate this enzyme in cartilage. The alternate explanation is decreased synthetic capacity with age. Palmoski et al. [6] have suggested that decreased binding capacity of the central core of hyaluronic acid might be an early step in degradation. Lysozyme may have a role in the breakdown of link protein. Other enzyme activities have been suggested.

In the excellent study by Santer, White, and Roughley [3] comparing aging cartilage from the tibial plateau, it was found that none of the marked chemical changes of osteoarthritis were found in the areas of superficial degeneration except for increases in keratosulfate ratios and the sialic acid–uronic acid ratios, both consistent with increased activity of cartilage proteases such as cathepsins. Because of the unavoidable lack of appropriate controls, they were unwilling to draw final conclusions from their study. They did suggest that as long as the superficial zone, though degenerate, was biomechanically intact, changes in the deeper zones were few. A similar situation is seen in the so-called odd facet of the patella, as described by Meachim and colleagues [7]. In contrast, there is the rather enigmatic syndrome of chondromalacia of the patella in adolescents and young adults, which exhibits softening and fibrillation of the deep layers often seen after various sorts of trauma and which could have a genetic factor as well. Modern biomechanical testing of young normal, aging, and degenerating cartilage has recently been done by Armstrong et al. [8] to show that experimental direct impact or shearing stress may induce horizontal clefts in the deep layers of cartilage and lead to degeneration. Meachim et al. [7] have described such clefts in human material.

We [9] have recently done similar experiments, simulating a moderate dashboard injury to the knee with a weighted pendulum. The force was less than that required to cause fractures. Our aim was to see what connection there was between mechanical trauma and biochemical degradation. We found a 400+ percent increase in the synthesis of arachidonic acid in cartilage cell membrane phospholipids. Since arachidonic acid is the major precursor of prostaglandins, we offer a cascade hypothesis of the pathway of catabolism in cartilage (Fig. 7-1). This is an expanded "final common pathway" of degradation. It is also obvious that there is a pathway of matrix synthesis and that in normal cartilage the two are in balance. Fulkerson et al. [7] have tested cartilage slices from young and old dogs, rabbits, and pigs and find that articular injections of arachidonate and prostaglandin E have little or no effect on young cartilage, whereas in older animals matrix degradation results. Interesting in this regard is that growth hormone reversed the effect of experimental damage to cartilage [11]. Similar healing effects have been noted with somatomedin by Sledge [12] and with uridine diphosphate by Erlich et al. [13]. It is clear enough that aging cartilage has less capacity to repair its matrix than does young cartilage.

Figure 7-1

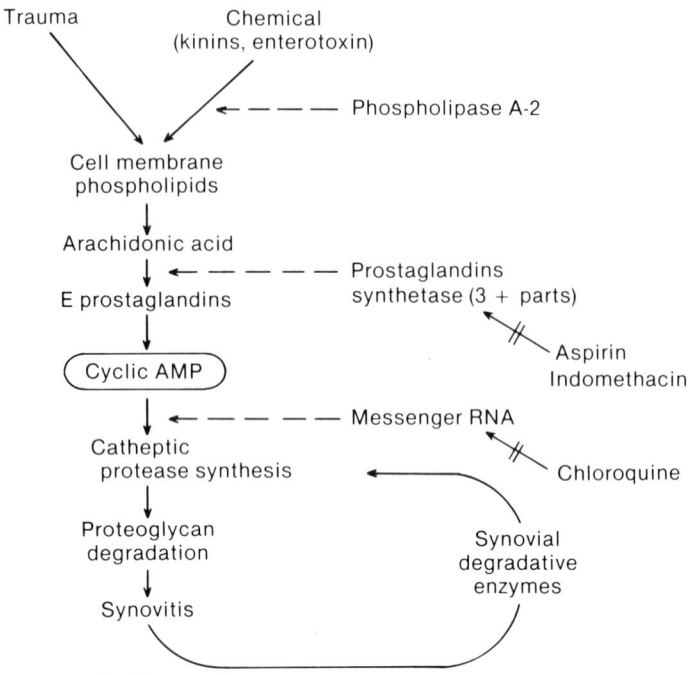

```
Trauma          Chemical
   \         (kinins, enterotoxin)
    \           /
     \         /
      \       / ←----- Phospholipase A-2
       \     /
        \   /
     Cell membrane
      phospholipids
         ↓
     Arachidonic acid
         ↓  ←----- Prostaglandins
     E prostaglandins    synthetase (3 + parts)
         ↓                    ↖
                               ╫ Aspirin
     (Cyclic AMP)              Indomethacin
         ↓  ←----- Messenger RNA
     Catheptic                  ↖
     protease synthesis          ╫ Chloroquine
         ↓                ←
     Proteoglycan            Synovial
     degradation             degradative
         ↓                    enzymes
     Synovitis
```

The osteoarthritic cascade.

CONCLUSIONS

Now let us return to our original questions. What are the effects of age on cartilage? Answer: surface degeneration in some areas, with evidence of a mild increase in proteolytic activity and strong evidence of a decreased capacity for repair. Is there any connection between aging and osteoarthritis? I believe there is a strong connection, on the basis of metabolic balance. Maintenance of cartilage function requires that synthesis can balance degradation. In *David Copperfield* Charles Dickens through Mr. Micawber wrote, "Income: twenty pounds, expenditures: twenty pounds less sixpence, result—happiness. Income: twenty pounds, expenditures: twenty pounds and sixpence, result—misery."

When the ability to synthesize new matrix replacements is diminished, lesser traumas and other insults are not easily repaired. The negative balance leads to cartilage softening, fibrillation, and the other changes of osteoarthritis. Once the 50-50 balance has been lost, degeneration and painful synovitis occur and can lead to misery and total joint replacement. There is an interesting dividend from this line of thought that concerns chondromalacia of the patella. Chondromalacia looks like osteoarthritis but does not act like it. The stress young people put on their kneecaps is formidable. It is very likely that horizontal clefts deep in the cartilage as described by Armstrong et al. [8] and Meachim et al. [7] are produced; certainly fibrillation and synovitis are common. And yet, with restriction of overactivity and by other conservative measures, a satisfactory clinical result can be obtained. What is the difference? Youth. Youth has what age lacks—a strong ability for matrix repair, possibly based on growth hormone or other reparative factors. Chondromalacia is not a separate entity but simply osteoarthritis in young people.

In summary, there is not great difference between the underlying processes of aging and osteoarthritis. The results are a matter of metabolic balance, synthesis versus degradation.

REFERENCES

1. Crelin, E.S., Southwick, W.O. Mitosis of chondrocytes induced in the knee joint cartilage of rabbits. *Yale J. Biol. Med.* 33:243, 1960.
2. Johnson, L. Biomechanics of degenerative joint disease. *Bull. Rheum. Dis.* 17:948, 1958.
3. Santer, V., White, R. J., Roughley, P.J. Proteoglycans from normal and degenerate cartilage of the adult human tibial plateau. *Arthritis Rheum.* 24:691, 1981.
4. Von Kuhn, R., Leppelmann, J.J. Galaktosamin und Glucosamin im Knorpel im Abhangigkeit von Lebensalter. *Liebigs Ann. Chem.* 611: 254, 1958.
5. Bollet, A.J. An essay on the biology of osteoarthritis. *Arthritis Rheum.* 12:152, 1969.
6. Palmoski, M.J., Colyer, R.A., Brandt, K.D. Joint motion in the absence of normal loading does not maintain normal articular cartilage. *Arthritis Rheum.* 23:325, 1980.
7. Meachim, G., Denham, D., Emery, I. H., Wilkinson, P.H. Collagen alignments and artificial splits at the surface of human articular cartilage. *J. Anat.* 118:101, 1974.
8. Armstrong, C.G., Schoonbeck, J., Jr., Moss, G., Mow, V.C. Characterization of impact-induced microtrauma to articular cartilage. *Trans. Am. Soc. Med. Engineers* 18:20, 1980.
9. Chrisman, O.D., Ladenbauer-Bellis, I.M., Panjabi, M. The relationship of mechanical trauma and the early reactions of osteoarthritic cartilage. *Clin. Orthop.* 161:275–284, 1981.
10. Fulkerson, J.P., Damiano, P., Williams, L., Chrisman, O.D. Studies of adult and young dog articular cartilage depletion by prostaglandin E in tissue culture. *Trans. Orthop. Res. Soc.* 5:282, 1981.
11. Chrisman, O.D. The effect of growth hormone on established cartilage lesions. *Clin. Orthop.* 107:232, 1975.
12. Sledge, C.G. Growth hormone and articular cartilage. *Fed. Proc.* 32:1503, 1973.
13. Ehrlich, M.G., Mankin, H.J., Treadwell, B.V. Uridine diphosphate (UDP) stimulation of protein-polysaccharide production. *J. Bone Joint Surg.* [Am] 56A:1239, 1974.

The Biology of Aging Bone

▶

Howard Duncan, M.D.
A. Michael Parfitt, M.B.

From the time of skeletal maturity, most people lose bone progressively with increasing age. This is such a common occurrence that it must be regarded as normal, and it becomes rather arbitrary to determine where a deviation from normal exists. However, there is no doubt that morbidity has a close relationship to the amount of bone present. For these reasons, attempts at quantitating total bone mass and measuring bone loss have been made. Considering the numerous methods available for measuring skeletal mass, the results published so far are often embarrassed by the fact that they are cross-sectional studies rather than longitudinal evaluations and thus merely represent the dispersal in society of those individuals whose bone mass may or may not be normal. It does not clearly recognize that the society's nutritional state may have changed during the life span of the older individuals and that younger persons presently being evaluated have improved their bone mass by reason of an altered dietary pattern or exercise. Hence, the longitudinal studies of the younger generation might not match the present cross-sectional studies of preceding generations at any specific age. Some methods used for assessment of bone mass include

1. Radiographic density of finger bones
2. Vertebral radiographic densitometry
3. Photon absorptiometry
   a. Photon absorptiometry across the radius and ulna
   b. Dual photon absorptiometry for inaccessible bones, vertebrae

The authors wish to recognize the assistance of C.H.E. Mathews and A.R. Villanueva in preparing material for this manuscript.

4. Total body calcium estimates
5. Density of iliac bone plugs
6. Mineral content of femoral fragments
7. Histomorphometric and microradiographic studies
8. Metacarpal cortical thickness measurements
9. Quantitative computerized tomography

RACE AND SEX

Bone loss has been shown to occur with age in all populations so far examined, and it appears to be a universal and inevitable human characteristic [1]. The magnitude of this bone loss differs in amount and rate between different ethnic groups. The South African Bantu women lose significantly less bone with age than white women who live in the same area and whose diet is significantly higher in calcium [2]. The same ethnic differences are also apparent in comparable groups found within the United States: bone loss is less severe in blacks than in whites [3, 4].

Bone loss begins earlier and is greater in women than in men—a difference that is temporally more clearly related to the menopause than to the age of the individual [5]. While the mean rate of bone loss in any given population is approximately the same for different bones and continues at a relatively constant rate with increasing age after maturity, there are, however, some differences in the pattern and distribution of the loss between certain individuals when identified in prospective longitudinal studies. Some individuals may lose no bone at all for some period, even up to 10 years [6], and others will show a more rapid bone loss. In general, women achieve their maximum bone mass between the ages of 20 and 30, and thereafter a mild bone loss occurs until the rapid demineralization during the first 5 postmenopausal years, after which time the rate of loss decreases. In men, maximum bone mass is achieved again between 20 and 30 years of age, and it would appear that the loss of bone mass with age occurs at approximately 1 percent per year on a relatively constant basis (Fig. 8-1). The female bone loss, after the immediate accelerated change due to estrogen withdrawal associated with the menopause, slowly approaches the same rate of loss as men [7].

THE MECHANISM OF BONE LOSS

Since it is recognized that there is no significant change in chemical analysis of bone mineral with aging, it becomes even more relevant to understand the mechanism by which

Biology of Aging Bone

bone is actually lost, for this also provides an understanding of the mechanisms of *excessive* bone loss leading to the osteopenic states.

The amount of bone present in the body at the time of skeletal maturity depends on the accumulation of bone dur-

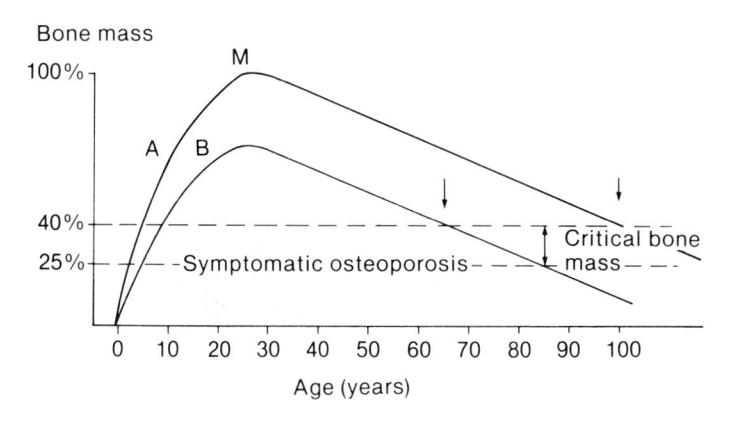

Figure 8-1

Osteoporosis in relation to maximal bone mass at maturity. M = maximal bone mass at age 20–25. A = optimal development of skeleton and physiologic rate of bone loss with age. B = suboptimal development of skeleton and physiologic rate of bone loss with age. ↓ = onset of symptomatic osteoporosis.

ing growth of both *trabecular* bone, resulting from enchondral ossification, and *cortical* bone, resulting from the net periosteal and endosteal apposition. If we assume that the same rate of bone loss with age occurs in the general population, then those persons with less bone present at the completion of growth will inevitably achieve lower levels of total bone mass sooner and be more likely to reach critical levels with structural failure and fracture than those who had more bone volume at their maturity (Fig. 8-1). In tubular bones, the cortical thickness depends on a balance between the net cellular activity on the periosteal surface with respect to bone formation and the net cellular effect on the endosteal (bone marrow) surface, with respect to bone resorption. The periosteal net apposition continues very slowly throughout life (Fig. 8-2); that is, the periosteal external diameter continues to increase even after cessation of normal growth and maturity. This phenomenon is recognized not only in the hands but also in the femur and skull. In the second metacarpal bone—that most usually identified for measurements—the periosteal diameter gains about 0.4 mm between the ages of 20 and 80, so the change may be difficult to demonstrate in any individual. In the femur, an overall periosteal expansion of 3.6 mm has been recorded by Smith and Walker [8].

The endosteal resorption also continues throughout life. During the growing years, rapid expansion of the bone marrow cavity keeps pace with the rapid periosteal apposition, for an imbalance between the two could lead either to osteopetrosis, where failure to expand the bone marrow cavity

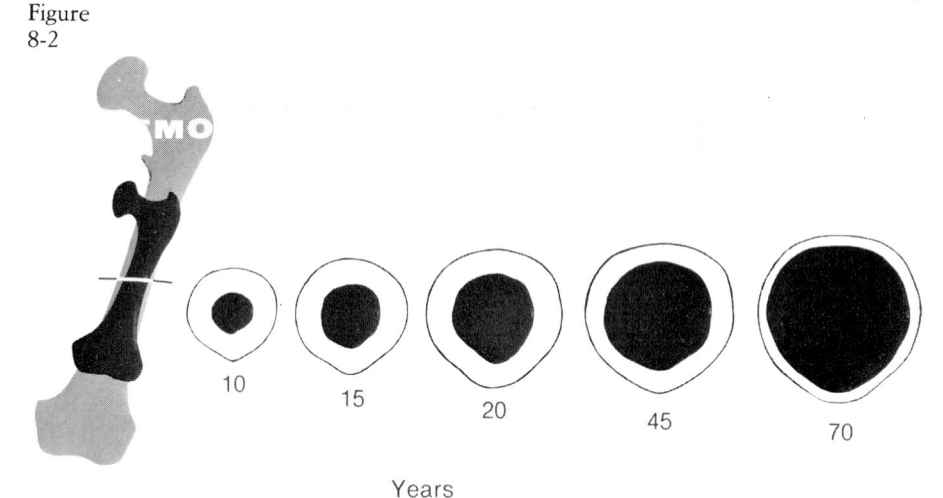

Figure 8-2

Years

Schematic representation of a cross-section of midfemur during growth, demonstrating the continuous expansion of the periosteal diameter and of the bone marrow cavity.
SOURCE: From Duncan, H., Jaworski, Z.F., Osteoporosis. In Spittell, J.A., Jr. (ed.), *Practice of Medicine*, vol. 5. Hagerstown, Md.: Harper & Row, p. 61.

occurs, or osteoporosis, where the cortical mass is lost due to the continuous and unbalanced erosive activity of the endosteum. In the aging process, the metacarpal again reflects the other tubular bones, where the magnitude of the endosteal change is greater than the periosteal expansion of 0.4 mm, for the medullary cavity increases by about 1.8 mm during the span of 20 to 80 years (Fig. 8-3). This slow and inexorable encroachment of the bone marrow cavity toward the periosteum results in the net reduction in cortical cross-sectional area of bone by about 25 percent. A similar loss of trabecular bone has been shown by morphologic measurement in iliac biopsy samples [9] (Mathews, Villanueva, Parfitt, Duncan, unpublished data), and in the vertebral bone at autopsy [10].

Together with these gross cortical dimension changes, a minor degree of intracortical porosity is seen to develop with age. In the femur, there is an increasing accumulation of large resorption spaces in the inner half of the cortex [11, 12] and there is also an increased number of partly filled resorption spaces, where haversian systems may complete only 30 to 50 percent of their original external dimension [13, 14]. In the upper extremity, the direct measurement of porosity is lacking, but indirect estimates of mineral density from combined studies with radiodensity and morphometry suggest that intracortical porosity is slight in the metacarpals but significant in the radius. However, when assessing the skeleton as a whole, the increase in intracortical porosity itself makes only a modest contribution to the total loss of bone with age.

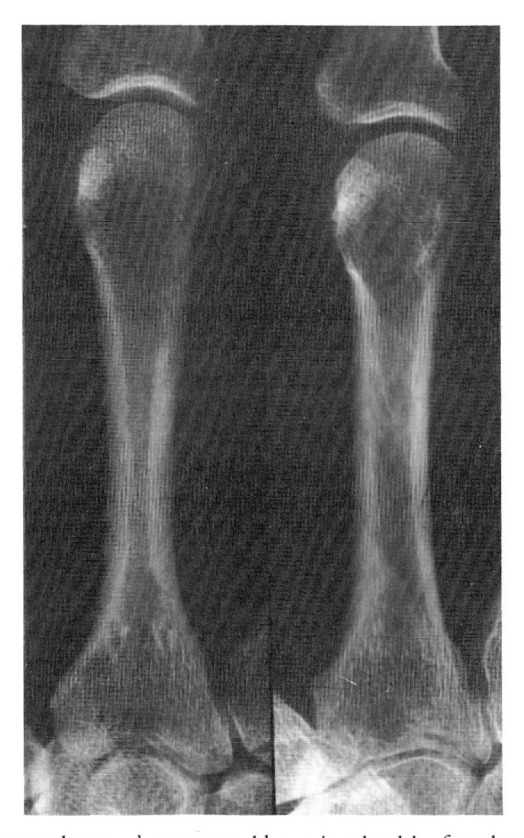

Figure 8-3

X-ray of a normal second metacarpal bone in a healthy female at the age of (A) 45 and (B) 70 years. Note the change in cortical thickness and the diameter of the bone marrow cavity at midshaft.

BONE REMODELING

The bone balance between net gain and net loss at each of the major bone surfaces (periosteum, endosteum, trabecular surface) is the result of a constant remodeling process taking place simultaneously at different speeds at all three sites. This remodeling and renewal of bone is thought to improve the mechanical resilience of bone and its structural alignment, to promote nutritional exchanges, and to repair microdamage that occurs during normal life. The remodeling process takes place in a discrete and circumscribed manner, involving a finite amount of bone removal and replacement at each site. This may be referred to as the quantum nature of the bone microanatomy. In standard cross sections of cortical bone, this unit of bone renewal is recognized as an osteon, through the center of which traverses a haversian canal (Fig. 8-4). In three-dimensional appreciation, the remodeling site will be compared with a tube some 150 to 250 μ in diameter,

Figure 8-4

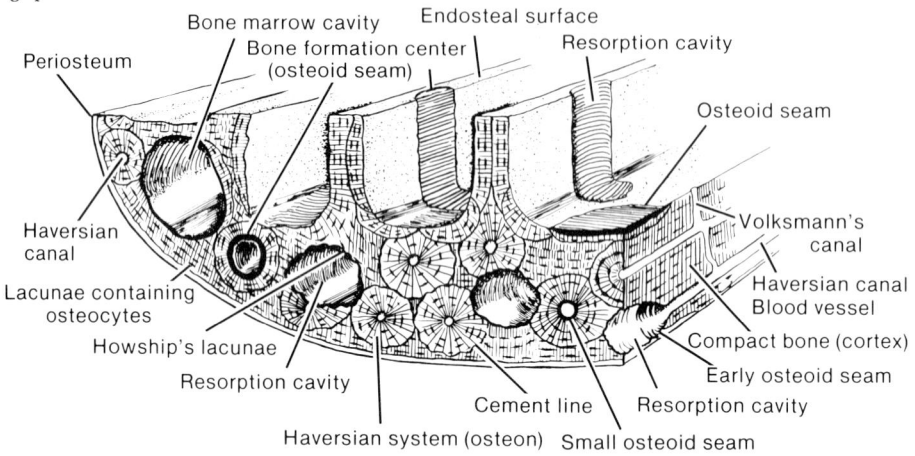

Cross-section and longitudinal section of the rib, showing the remodeling units in cortical bone and on endosteal surface.
SOURCE: From Duncan H., Jaworski, Z.F., Osteoporosis. In Spittell, J.A., Jr. (ed.), *Practice of Medicine*, vol. 5, Hagerstown, Md.: Harper & Row, p. 61.

drilled in a direction parallel to the longitudinal axis of the bone. In normal human subjects, the sequence of bone removal followed by formation takes approximately 3 to 4 months. At different ages, a certain percentage of such bone units is active, and approximately 2 to 4 percent of the total cortical bone is turned over in this fashion per year. A new unit may encroach on a previously completed unit or deviate to repair some slightly damaged or necrotic area, and it is quite evident from this system that the skeleton itself resembles a building made of individual bricks rather than one made of poured concrete. While this feature has been recognized for more than a century, its implications for the quantitation of bone remodeling were first appreciated by Frost in 1963 [15]. Also with increasing age, the anatomic distribution of haversian canals in the cortices is altered, for there is an increasing number of overlapping osteons, and partial osteons, remnants, and interstitial bone will increase in amount. These areas increase not only in their number with age but also in their degree of mineralization [16].

Several other changes with age have also been recorded. There is an increased frequency of plugging and calcification of the osteocyte lacuna, and the canaliculi as well as osteons themselves may become occluded and subsequently mineralized [17, 18]. When bone ages and osteocytes wither and die, the bone around the lacunae becomes hypermineralized; this has been referred to as micropetrosis. This event indicates that the osteocyte was partly effective in preventing complete saturation with mineral during its normal life cycle insofar as it held the level of mineralization at 10 percent less than the maximum. Micropetrosis is accompanied by increased brittleness, decreased mechanical strength, and increased susceptibility to fatigue and microfracture. All these changes are more evident in the extrahaversian bone than in the haversian systems.

Many of these changes cannot be recognized in decalcified sections. Frost [15] emphasized the cellular sequence in repair of such sites. He showed that remodeling took place in three different stages: (a) *activation,* where a primary stimulus promoted the generation of preosteoclast cells at the site requiring repair, (b) a phase of osteoclastic *resorption* or removal of bone, and (c) osteoblastic repair with new bone *formation* at the previously excavated site. This sequence of events is readily appreciated in the cortical bone where the osteon is the completed unit, but less obvious at the periosteal, endosteal, and trabecular surfaces. In these latter sites, the first step of the remodeling sequence is the excavation by a team of newly formed osteoclasts of a cavity of about 50 μm deep, a performance that takes approximately 1 to 2 weeks. When the osteoclasts have finished their work, they are replaced by transitional cells of uncertain nature, which deposit the cement substances. This cementing line is more radiodense than the surrounding bone and excludes any communication or mineral diffusion between adjacent osteons, for it interrupts canalicular as well as collagen fiber continuity. Following this cementing period of a few days, a team of new osteoblasts appears and refills the cavity with new bone. This performance takes an extended period of 2 to 3 months in the 4-month cycle. This complete remodeling period, activation-resorption-formation, is referred to as the sigma value peculiar to each of the mammalian species and to different individuals at different times. The rate of *total* bone turnover depends on the *frequency* with which *new* bone remodeling units are initiated and completed and *not* specifically on the rate at which individual cells work or on the size of these individual units. In the skeleton with a turnover rate of 10 percent per year, one remodeling unit is initiated and one completed every 10 seconds.

With age, bone is lost from the inner surfaces of the

cortices, which become progressively thinner; however, there is simultaneous loss from both surfaces of many individual trabeculae. This dual involvement of trabeculae may not only thin such bone but also actually perforate and cause complete disappearance of numerous trabeculae. By the age of 75, the average woman has lost about 25 percent of her cortical bone and about 40 percent of her trabecular bone. In applying this quantum nature of bone remodeling to the event of thinning of cortices, it is to be recognized that the net loss of bone represents a net excess of resorption over bone formation at each unit. This could result from (a) an abnormally deep excavation being developed by osteoclasts with a normal volume of bone replacement (albeit an incomplete refilling), (b) normal resorption volume with a failure to replace all the bone excavated, or (c) a less than normal excavation with an even smaller refilling. At different stages of disease, all of these aspects have been identified, but in the aging process, this local imbalance of bone remodeling is due to a normal or slightly reduced size of excavation followed by an unmatching decrease in osteoblastic activity. At the periosteal surface, of course, where expansion continues, the opposite applies, such that each remodeling cycle puts back slightly more bone than it had removed.

By the use of tetracycline labeling methods, it has been possible to quantitate many aspects of bone remodeling, for this drug is incorporated into newly deposited bone mineral at the junction between mineralized bone and the osteoid. Sites of bone formation can be made visible by the yellow fluorescence of tetracycline with ultraviolet or deep blue light. By these means, it is possible to distinguish between a change in the *number* of bone-forming *cells* and the *number* of *units,* osteons, or haversian systems, as well as a change in the activity of individual bone-forming cells, osteoblasts. However, tetracycline labeling cannot by itself distinguish between two mechanisms of net bone loss (i.e., osteoclast holes too deep or osteoblast repair too shallow) because net balance at the *completion* of each cycle of structural renewal depends not on the speed of the cells but on their *stamina* or, more precisely, on the total amount of work performed. When evaluating bone remodeling sites using tetracycline labels, one does not completely accommodate the observation that haversian systems might only be one-half or two-thirds completed until after a succession of remodeling cycles has been completed. Also, at the time of bone biopsy of an osteoporotic subject, the current tetracycline labeling data may represent a dynamic state long after the period when the maximum loss occurred.

Measurements of the mean wall thickness of the haversian canal or, on trabecular surfaces, the distance between the surface and the cementing line remain constant until about the age of 50 and then decrease with increasing age with no significant difference between the sexes. The osteoblasts evidently have less stamina after this age and stop making bone before the resorption cavities are completely refilled. There is also evidence to show that there is progressive reduction in depth of resorption cavities, that is, they are shallower than previously; this would indicate that osteoclasts also lose stamina with age, but not so markedly as osteoblasts. In the repair of bone, the osteoblastic activity is maximal during the first 2 weeks of life span of the repairing osteoblasts, when approximately 70 percent of the bone matrix is replaced. Interference in this pattern of repair is recognized in patients with very severe reduction in bone turnover and with increasing age when the initial rate of matrix synthesis is reduced, and the osteoblasts stop making bone matrix altogether when their task is only about two-thirds complete. An exaggerated phenomenon but of a similar type is seen in patients with a severe symptomatic osteoporosis when compared with subjects of the same age.

CONCLUSIONS

It can be stated with certainty that age-dependent bone loss is a universal observation that is present in all humans but varies at different times in the two sexes and at different rates in different ethnic groups. In some, these same phenomena lead to the disease of involutional osteoporosis, characterized by fractures, bone pain, and loss of height. This clinical state depends not only on the reduction in the amount of bone but also on its quality. The former is a quantitative departure from normal. The amount of bone present at the onset of clinical disability is a function of the amount present at the time of skeletal maturity and the rate of bone loss with age. Qualitative changes in bone that may increase the risk of clinical effects include alterations of the shape, reduction in mechanical strength due to redistribution of trabecular bone, and retarded healing of microfractures.

With aging, the bone composition of collagen, fat, calcium, and mineral shows no quantitative differences between men and women. In older people, there seems to be a spectrum of bone status; at one end are the light-boned osteoporotic women without osteoarthrosis who suffer fractures, and at the other are individuals with strong bone without fractures who suffer from osteoarthrosis [7]. Evidence is accumulating that suggests that, in addition to race and sex, there are subsets of individuals whose aging processes of bone occur at different rates at different times, and that these changes may be intermittent or continuous.

REFERENCES

1. Garn, S.M. *The Earlier Gain and the Later Loss of Cortical Bone.* Springfield, Ill.: Charles C Thomas, 1970.
2. Dent, C.E., Engelbrecht, H.E., Godfrey, R.C. Osteoporosis of lumbar vertebrae and the calcification of abdominal aorta in women living in Durban. *Br. Med. J.* 4:76–80, 1968.
3. Smith, R.W., Rizek, J. Epidemiological studies of osteoporosis in women of Puerto Rico and Southeast Michigan with special reference to age, race, national origin, and other related or associated findings. *Clin. Orthop.* 45:31–48, 1966.
4. Bollet, H.A. Epidemiology of osteoporosis. *Arch. Intern. Med.* 116:191, 1965.
5. Nordin, B.E.C., Young, M.M., Denilly, B., et al. Lumbar spine densitometry, methodology, and results in relation to menopause. *Clin. Radiol.* 19:459–464, 1968.
6. Adam, P., Davies, G.T., Sweetnam, P. Osteoporosis and the effects of aging on bone mass in elderly men and women. *Q. J. Med.* 39:601–616, 1970.
7. Dequeker, J. Bone and aging. *Ann. Rheum. Dis.* 34:100–115, 1975.
8. Smith, R.W., Walker, R.R. Femoral expansion in aging women. *Science* 145:156–157, 1964.
9. Bordier, P.J., Tun Chot, S. Quantitative histology of metabolic bone disease. *Clin. Endocrinol. Metab.* 1(1):197–215, 1972.
10. Arnold, J.S., Bartly, M.H., Taunt, S.A., Jenkins, D.P. Skeletal changes in aging and disease. *Clin. Orthop.* 49:17–38, 1966.
11. Atkinson, P.J., Weatherell, J.A., Weidman, S.M. Changes in density of human femoral cortex with age. *J. Bone Joint Surg.* [Br] 44B:496–502, 1962.
12. Jowsey, J. Age changes in human bone. *Clin. Orthop.* 16:210–217, 1960.
13. Urist, M. Accelerated aging and premature death of bone cells in osteoporosis. In Pearson, O.H., Joplin, G.F. (eds.). *Dynamic States of Metabolic Bone Disease.* Philadelphia: F.A. Davis, 1964, pp. 127–155.
14. Jaworski, Z.F., Meunier, P., Frost, H.M. Observations on two types of resorption cavities in lamellar cortical bone. *Clin. Orthop.* 83:279–285, 1972.
15. Frost, H.M. *Bone Remodelling Dynamics.* Springfield, Ill.: Charles C Thomas, 1963, p. 61.
16. Jowsey, J. Aging of bone. *Clin. Orthop.* 17:216, 1960.
17. Tomes, J., DeMorgan, C., 1853. Quoted by Enlow, D. H., Functions of the Haversian system. *Am. J. Anat.* 110:269–305, 1962.
18. Frost, H.M. In vivo osteocyte death. *J. Bone Joint Surg.* [Am] 42A:144–150, 1960.

Osteoporosis

▶

Jenifer Jowsey, D.Phil.

9

DEFINITION OF OSTEOPOROSIS

There are few diseases that afflict one sex more frequently than another, but osteoporosis is one of them; women form the large majority of the group who seek medical help for this disease. The most common complaint is pain, usually leading to some degree of disability, and in this sense osteoporosis is a true disease, or "dis-ease." The cause of the pain is the occurrence of fractures of the bone, either present on an x-ray of the painful area or as microfractures. Both kinds will cause compression of nerves and, therefore, muscle spasms and pain. In some areas of the skeleton, such as the upper end of the femur, a "hip fracture" will cause mechanical instability and thus significant disability.

The fractures are typically found in three areas, the lumbar and thoracic spine, the neck of the femur, and the distal forearm—the so-called Colles' fracture. These are all areas of the skeleton subject to high loads. The lower part of the spine is an area where a comparatively small volume of bone tissue supports the entire upper half of the body. The vertebrae, which form the spine, consist largely of spongy or trabecular bone with only a narrow rim of cortical bone forming the outer wall of each vertebral body. The amount of stress on a unit area of bone in vertebrae is therefore high and is one reason why vertebrae are frequently the first site for the development of fractures in osteoporosis. Because most people tend to lean forward and bow their backs when sitting down to read, write, or eat, the vertebrae of the spine often undergo anterior wedging or an incomplete fracture through the ventral part of the body of the vertebrae. This is not different from a symmetric vertical fracture except in the extent of bone collapse.

The neck of the femur is also an area where a small volume of bone supports a comparatively large amount of weight. The neck is at an angle to the vertical stress imposed on it, and although the trabecular bundles are distributed in relation to the lines of stress, the external cortex is at an angle to the vertical. Colles' fractures of the forearm are often directly related to a fall on the extended arm, which results in fracture of the radius and ulnar just proximal to the wrist.

Even a cursory consideration of the site of fracture in osteoporosis and the structure of the skeleton will show that the fractures in this disease are the result of a combination of a decreased bone mass and stress or strain. A person with osteoporosis is therefore one with a decreased amount of bone who has sustained minor trauma or stress and sustains a fracture, almost invariably at one of the three sites mentioned. The difference between a patient with osteoporosis and a "nonpatient" lies in the existence of a fracture in the former while the nonpatient merely may not have experienced the trauma needed to cause a symptomatic fracture. Quantitative measurements of bone mass reveal a wide range of values for men and women of different ages; patients with symptomatic osteoporosis may have a total bone mass that is indistinguishable from that of people without osteoporosis. The disease is diagnosed clinically by the presence of fractures that have occurred with only minor trauma.

Radiologic Appearance of Osteoporosis
An x-ray of a patient with osteoporosis will, of course, demonstrate the fracture and will also show generalized loss of bone in the rest of the skeleton. The cortices will be thin, and in bones that normally have a thin cortex, such as the ribs and clavicles, the x-ray will show "penciling" or a narrowing of the cortex to a single line. Trabeculae will be sparse so that the remaining trabeculae stand out. Bone is lost in inverse proportion to the importance of the structure for support; in other words, the bone of least structural importance is lost first. Therefore, the outer cortex and the vertical trabeculae are preserved; indeed, they are almost never lost and remain even in patients who have had osteoporosis for a long time. Bone loss appears to slow in long-term patients, and the disease burns out.

Because bone is lost preferentially from the structurally less important areas, a measure of the disappearance of trabecular bone can be used to estimate the total bone loss that has taken place in an individual. The femoral trabecular pattern index has been used to evaluate bone loss. It was first described by Singh and is usually referred to as the Singh index [1]. The process of trabecular loss in the upper end and head of the femur is a continuous process; however, Singh recognized seven stages. In the first (stage VII), where no loss of bone has occurred, the x-ray of the hip shows a continuous mass of trabeculae. The minor trabeculae disappear first and reveal the three major bundles of trabeculae to give a Singh index of VI. Further loss results in the disappearance of these three bundles consecutively until only the major compressive bundle is left at a Singh index of I. While this method is subjective and inaccurate in that it imposes a series of numbers on a continuous process, it has proved useful for two reasons. First, it requires a simple anterior-posterior x-ray of the hips, which can be read by a radiologist with a little experience; second, and most important, it evaluates the amount of bone that has been lost rather than the absolute amount of bone. Methods of measuring bone *mass* may be more accurate but are often irrelevant to the presence or absence of bone *loss*; for example, a large man who should have a large skeleton may lose bone and be a candidate for fracture and yet have a bone mass that is normal for a slight man. It has not been useful to correct bone mass values by height, weight, or surface area to obtain a value representing the amount of bone present in an individual as compared with the amount they should have. The Singh index is the only method that provides this information. A Singh index of III, II, or I is almost always present in patients with osteoporosis, while nonosteoporotic patients have a Singh index of VII to IV.

Recently, computerized axial tomography has been developed for accurate bone density studies and is particularly valuable for longitudinal studies [2].

Types of Osteoporosis

A number of clinically defined disorders are associated with bone loss and osteoporosis. These can be differentiated from other forms of osteoporosis because of related abnormalities.

Disease such as hyperparathyroidism, hyperthyroidism, and hypercortisonism are associated with osteoporosis that occurs throughout the skeleton. Other types of osteoporosis may be localized to a specific part of the skeleton. Disuse osteoporosis occurs in areas of bone that are immobilized in a cast or by other means. Denervation also causes localized bone loss.

These two types of osteoporosis—endocrine-related and disuse—are generally clearly attributable to a single, easily defined cause. In this respect they are different from the generally multicausal osteoporosis discussed in this chapter.

The factors found as causative in the majority of patients are
1. Low calcium content of the diet
2. High phosphorus content of the diet
3. Lack of estrogen

The osteoporosis caused by these factors is usually referred to as idiopathic, postmenopausal, or senile.

Idiopathic osteoporosis occurs in people between the ages of 20 and 50. Postmenopausal osteoporosis occurs in women who are postmenopausal, including those who have undeveloped ovaries (Turner's syndrome) or gonadal dysgenesis and those who are producing decreased amounts of estrogen. Senile osteoporosis occurs in men and women of advanced years, and juvenile osteoporosis is found in children and adolescents.

All the causes listed are important to a greater or lesser degree in osteoporotic patients with few exceptions. In other words, a woman of 65 years will have lost bone because she is postmenopausal, calcium deficient, has a high phosphorus intake, and is relatively inactive. The exception would be a child, who is, of course, not estrogen deficient and in whom calcium and phosphorus imbalances are the major cause of the osteoporosis. For this reason it is often hard to define in any one patient a single cause for the disease, although one cause may appear and indeed may be predominant.

The Pathology and Physiology of Osteoporosis

The skeletal x-rays of a patient with osteoporosis show bones that are thinner and contain fewer trabeculae than normal. A piece of bone from such a patient will show a narrow cortex, porous cortical bone, and a decreased amount of spongy or trabecular bone. Bone loss can occur as a result of an increased amount of resorption of bone, a decreased amount of formation of bone, or both. Comparisons of bone formation and resorption in osteoporotic and nonosteoporotic people have shown that, in osteoporosis, resorption is increased, while new tissue formation may be normal or decreased if inactivity is a major factor in the cause of bone loss. There is no instance of osteoporosis in which bone formation is decreased and resorption is normal.

The two types of cells responsible for resorption and formation of bone are the osteoclasts and osteoblasts. The osteoclasts are derived from blood-borne cells, are multinucleate, and destroy the mineralized matrix of the tissue. The osteoblasts have a single nucleus, are derived from the fibroblast line of cells, and produce the matrix of bone. The minerals, calcium and phosphorus, are deposited in the matrix later. A number of different factors affect the metabolism of these cells and, as a result, will increase or decrease bone loss (Table 9-1).

Increased bone resorption in osteoporosis is found throughout the skeleton and, therefore, the increased porosity also occurs throughout the skeleton. Because the level of

Table 9-1

Factors that Increase or Decrease Bone Loss and Resorption

Stimulate Bone Resorption and Cause Bone Loss
   High phosphorus
   Low calcium
   Acidosis
   Inactivity
   Steroids

Depress Bone Resorption and Slow Bone Loss
   Estrogen
   Alkylosis
   High-calcium diet
   Activity
   High ratio of calcium to phosphorus

bone turnover is lower in some areas of bone, such as the cortical bone, the rate of resorption of tissue, and thus the development of significant porosity, also tend to be lower. However, areas where the osteoporotic process is most evident on x-ray are not necessarily areas where a high turnover rate occurs but rather are where the amount of cortical bone is low, as in the vertebrae, the clavicles, or the ribs. The rate of loss of trabecular or spongy bone, in number of grams lost per week or year, tends to be lower than that of cortical bone, although the percentage loss is high because the mass of bony tissue is small in this type of bone. It is a mistake to believe that trabecular or spongy bone is particularly susceptible to the osteoporotic process. Although areas of the skeleton such as the spine do indeed contain a large proportion of trabecular bone and are a site for pathologic fractures, other areas of the skeleton, such as the distal tibia, contain just as much or more trabecular bone but are protected from fracture by the size of the bone or by biomechanical characteristics.

CAUSES OF OSTEOPOROSIS

From the discussion so far, it is evident that loss of bone results from increased resorption of the bony tissue and that the fracture is the consequence of the bone loss and increased stress in the fracture area. As shown in Table 9-2, the factors that play a role in bone loss are many; inactivity, calcium deficiency, phosphorus excess, and estrogen deficiency are clearly associated with the process and probably account for most of the bone loss that occurs in osteoporosis.

Table 9-2

Factors that Contribute to Osteoporosis

Calcium deficiency
High-phosphorus diet
Bed rest and decreased levels of activity
Estrogen deficiency
Fasting or dieting
High alcohol consumption
Steroid administration
Anticonvulsant therapy
Gastrectomy

The Importance of Maximal Bone Mass

After growth stops, the maximal bone mass is reached at about 20 years of age. From this age onward there is a gradual decrease in bone mass in both men and women. In this context the distinction between postmenopausal and senile osteoporosis becomes blurred in elderly women because, with time, both age- and estrogen-related bone loss are taking place. However, the important contribution of the maximal bone mass fits into the picture of osteoporosis at this point. Obviously, if fractures take place when a certain amount of bone has disappeared, and if all people are losing bone, then it is only a matter of time before all people will become osteoporotic if they live long enough. In this context, therefore, the greater the bone mass at age 20, the less chance there is of developing clinical osteoporosis with a fracture before death intervenes. Data indicate that, in general, men have a greater maximal bone mass at maturity than do women, and that black people have more bone mass than white people. Independently of the rate of bone loss, therefore, a black man will be less likely to develop osteoporosis than will a white woman.

There is not a great deal that can be done to alter ones sex or race; however, it is important to reach adulthood with as much bone as possible. Little attention has been paid to calcium requirements in children and adolescents, and it is likely that most young persons in the United States are calcium deficient. There is histologic evidence showing that bone porosity is higher between the ages of 10 and 17 than in the 20s. Changes in dietary habits, which include an increased consumption of soft drinks and snack foods, suggest that adults today achieve less than maximal bone mass at maturity. Calcium supplements would prevent the deleterious effects of the contemporary low-calcium, high-phosphorus diet. After maturity, it is the rate of loss that will determine whether a person will lose bone to the point at which fractures are likely.

Inactivity as a Cause of Bone Loss

A clear relationship has been established between bone mass and degree of activity. Increased stress on bone stimulates new bone formation, and lack of stress results in decreased formation and increased resorption. Exactly how stress modifies the behavior of bone cells is not really known but may be related in some way to the piezoelectric effect of stress on crystals. In the case of the skeleton, weight-bearing movement will produce the stress, while the collagen acts as the crystal that transforms the stimulus into an osteoblastic effect.

Studies have documented increased bone mass in professional athletes as compared with nonathletes, while bed rest is a well-recognized contributor to osteoporosis. In the average population, bone mass is generally higher in men who have an active occupation, such as farming or other forms of manual labor, while sedentary workers, such as office clerks and most professionals, have accelerated bone loss by comparison.

Calcium Homeostasis and Osteoporosis

The Control System in the Body The primary role of the skeleton is to supply calcium so that a normal level of calcium can be maintained in the serum. Ionic calcium is necessary for nerve conduction, cell membrane activity, blood coagulation, and muscle contraction; a subnormal serum calcium level will rapidly cause death. The maintenance of normal calcium is achieved by a close interplay between the level of ionized calcium in the blood, the parathyroids, and osteoclastic bone resorption. The parathyroids are sensitive to changes in the blood calcium level, increasing secretion of parathyroid hormone as the blood calcium level falls and decreasing secretion as the level rises. Parathyroid hormone circulates in the blood, where it acts on two target tissues, kidney and bone. In the kidney it increases the reabsorption of calcium, therefore decreasing excretion; and also increases the hydroxylation of 25-hydroxyvitamin D to 1,25-dihydroxyvitamin D, which in turn increases the intestinal absorption of calcium. The efficacy of these functions depends on the presence of calcium in the diet to a large extent. The action of parathyroid hormone on bone is to increase the activity of existing osteoclasts and cause new ones to differentiate so that osteoclastic bone resorption increases. Resorption of bone releases calcium into the blood to raise the calcium level.

Parathyroid hormone also affects phosphorus. Since bone contains phosphorus as well as calcium, these are both released when resorption occurs. The phosphorus is excreted by the kidney, and elevated levels of parathyroid hormone cause increased amounts of phosphorus in the urine. High levels of phosphorus in the blood will decrease renal hydroxylation of 25-hydroxyvitamin D, so the parathyroid-phosphorus effect in the kidney prevents cancellation of the calcium-conserving effect of the hormone. In general, the actions of parathyroid hormone are directed toward conserving normal levels of calcium in the blood and preventing high blood levels of calcium *and* of phosphorus. The latter would be a dangerous situation because of the tendency toward calcium phosphate precipitation in soft tissues whenever serum phosphorus and calcium levels are elevated. When kidney function is impaired and the renal action of parathyroid hormone is therefore inhibited, as in renal failure, the serum calcium–serum phosphorus product is high; soft-tissue calcification then becomes a problem.

Sources of Calcium Essentially the only sources of calcium in the diet are milk and milk products, such as cheese, butter, and ice cream. Other foods containing calcium are green leafy vegetables, such as cabbage, lettuce, and collards; however, the amount of calcium they contain is small, and these foods are not frequently eaten. During the last 20 years, a number of changes in eating habits that affect calcium intake have occurred in the United States, particularly among young people.

1. Consumption of soft drinks such as root beer and cola, which are low in calcium and high in phosphorus.
2. Consumption of fast foods and snack foods, such as crackers, hamburgers, potato chips, which are low in calcium and high in phosphorus.
3. Avoidance of cholesterol-containing foods, which are often those containing calcium, such as butter and cheese.
4. General increase in alcohol consumption: Any amount of alcohol inhibits the absorption of dietary calcium.
5. Calorie awareness. People avoid dairy products because of their generally high calorie content.

Osteoporosis

6. Repeated fasting [*or anorexia nervosa*]. Regular complete fasts for a period of a few days is not an uncommon practice and obviously results in zero calcium intake. Fasting also causes acidosis and more bone loss because bone acts as a buffer.

7. [*tobacco & caff*]

The Role of Calcium Deficiency Lack of calcium in the diet is a well-recognized contributing factor to osteoporosis. People who do not drink milk are more common in an osteoporotic population than in a nonosteoporotic population. Although 0.8 g is accepted as the minimum daily requirement for calcium, recent studies in perimenopausal women have shown that less than 1 g of calcium in premenopausal women will result in loss of bone, while postmenopausal women require 1.5 g of calcium per day [3]. Both values are well above the estimated average calcium intake, which several studies in populations of both osteoporotic patients and normal people have shown to be less than 0.8 g per day. A recent survey of the habitual calcium intake in postmenopausal women in the United States revealed a median value of 0.44 g; this lower value reflects the changes both in dietary habits discussed previously and in the food habits of older women.

If sufficient calcium is not available from a dietary source, this essential element must come from the skeleton. Large amounts of calcium in the food are reflected in the amount of calcium in the urine; for example, with a 1-g calcium supplement, urinary calcium may increase from 150 mg to 250 mg every 24 hours. However, even with little or no calcium intake, the amount of calcium in the urine will rarely fall much below 100 mg. Simple addition and subtraction will tell us that if 20 percent of a 440-mg intake is absorbed, 100 or more mg is lost in the urine, and a lesser amount in feces and sweat, then the average person is in negative calcium balance. Indeed, direct measurements of calcium balance indicate a negative balance, which becomes greater with increasing age [3].

The Role of Phosphorus in Bone Loss

Calcium deficiency as a cause of excessive bone loss is closely linked with phosphorus metabolism. The absorption of phosphorus from the diet is not closely controlled, and a high phosphorus intake is reflected in serum phosphorus levels and in the amount of phosphorus excreted in the urine. After a meal the serum phosphorus level rises by an average of 14 percent, which reduces ionized calcium and increases parathyroid hormone secretion and therefore bone resorption. A high serum phosphorus level is probably the most potent stimulus for the parathyroid glands. If renal function is impaired and phosphorus cannot be efficiently excreted, the serum phosphorus level may be markedly elevated. In renal failure destructive bone disease develops rapidly and is a major medical problem in such patients.

Almost every meal contains phosphorus, and the changes in dietary habits that have reduced calcium intake have increased phosphorus intake. Soft drinks, snack foods, weight-control diets, and processed foods all contain relatively large amounts of phosphorus. As the standard of living rises in the population, meat is eaten both more frequently and in large amounts, and meat, fish, and fowl all contain a high concentration of phosphorus. Processed food is becoming more popular; phosphorus is added to cheese, meat, and other natural products to soften them and make them more acceptable. The average phosphorus intake is approximately twice that of the calcium level of the diet.

The Calcium-to-Phosphorus Ratio

Phosphorus administration to animals produces bone loss, and in cats the loss of bone has been shown to progress to fractures—that is, symptomatic osteoporosis. Calcium deficiency has the same effect. In both instances, serum parathyroid hormone levels rise and bone resorption increases. Specific investigations sought to evaluate the relative importance of calcium deprivation and phosphorus excess in the bone loss and showed that it is the calcium-to-phosphorus ratio that is the critical factor in the stimulation of increased bone resorption [4]. If the amount of calcium is greater than the amount of phosphorus, resorption will not increase, but if there is less calcium than phosphorus, then bone resorption increases. Since most foods contain phosphorus and very few contain calcium without phosphorus, it would be virtually impossible to eat a diet with a calcium-to-phosphorus ratio greater than 1, unless intake were restricted entirely to milk and natural milk products, which most people would find unacceptable. However, calcium supplements are easy to get, and because it is the *ratio* of these two elements that is important, a calcium supplement is an efficient way of achieving a calcium-to-phosphorus ratio of greater than 1.

TREATMENT OF OSTEOPOROSIS

Fractures and Bone Loss

By the time a fracture occurs in osteoporosis, loss of bone from the skeleton is considerable. Fractures in osteoporotic individuals heal normally, and the discomfort and disability generally go away. However, further fractures will and do happen because the bone mass has decreased to an amount that is unable to support the individual. For osteoporosis to be treated successfully, the bone mass must be increased so that the amount is similar to that the individual had in youth. The alternative approach is to prevent the bone loss totally from occurring.

Many different forms of treatment for osteoporosis have been studied. The population is large and the disease is rarely lethal, so there are many patients who need treatment for many years. Because of the multicausal nature of the disease, the agents that have been considered as potentially helpful have been many and varied (Table 9-3).

Agents Considered to be Useful in the Treatment of Osteoporosis Table 9-3

Agent	Effect on Bone	Effective
Phosphate supplements	Increase resorption	No
Fluoride supplements	Increase resorption and osteoid production	No
Diphosphonates	Osteoid production	No
Calcitonin	Probably none	No
Growth hormone	Increases turnover	No
Estrogens	Depress resorption and formation	Partially
1,25-Vitamin D	Decreases resorption	Partially
Calcium supplements and calcium infusions	Decrease resorption	Partially
Fluoride and calcium	Decrease resorption and increase formation	Yes

Agents that Worsen Osteoporosis

Phosphates Oral phosphate supplements will reliably decrease the amount of calcium in the urine. Because this finding was associated with a more positive calcium balance, phosphate supplements were thought to cause a buildup of bone in the skeleton and at one time were considered as being of potential use in the treatment of osteoporosis. Many animal studies had been done that suggested that phosphorus in fact increased the holes in bone. Phosphorus supplements had been used in animal food to produce histologic osteoporosis in animals, which in some reports eventually developed into symptomatic fractures. Studies in animals are usually but not always an indication of what will happen in human beings, so a short investigation in a small group of osteoporotic patients was carried out to find out what happened in human beings [5]. As in animals, phosphorus supplements caused holes to appear in human bone cortex, accelerated the disappearance of spongy bone, and actually hastened the process of bone loss. These results were not surprising in light of the animal studies, and they are compatible with the role played by dietary phosphorus in causing osteoporosis.

The original observation that an increase in phosphorus intake lowers calcium excretion remains a consistent finding, although long-term animal studies showed that urinary calcium eventually increased. In animals it was also possible to show that the calcium that stayed in the body was deposited as calcium phosphate in soft tissues, especially the kidney and blood vessels. Therefore, aside from the deleterious effect on bone, phosphorus in amounts larger than those already occurring in the diet will cause soft-tissue mineralization.

Diphosphonates A class of compounds used industrially for the prevention of boiler-scale, the diphosphonates, were also studied as potentially useful agents in osteoporosis. Ethane-1-hydroxy,1-diphosphonate (EHDP), or sodium editronate, given to both osteoporotic and nonosteoporotic individuals produced an increase in serum phosphorus levels; examination of bone tissue revealed a marked increase in the amount of unmineralized bone. The increased osteoid or matrix resembled the condition of osteomalacia and disappeared slowly when the diphosphonate was stopped and calcium and vitamin D supplements were given. The high serum phosphate stimulated secretion of parathyroid hormone and, therefore, osteoclastic bone resorption and, like phosphorus supplements, accelerated bone loss.

Calcitonin Calcitonin, a recently discovered hormone that acts primarily on phosphorus, was also considered as a worsening agent in osteoporosis because it causes a decrease in serum calcium levels. This finding was interpreted to mean that the hormone was preventing calcium from coming out of bone by the process of resorption. In preliminary animal studies calcitonin had minimal effect on bone, and in later investigations in humans the bone changes were negligible, except in Paget's disease of bone. In Paget's disease, the rate of bone cell activity is very high in localized areas of the skeleton, and it may well be that calcitonin acts by depleting these cells of phosphorus and thus depriving them of the energy needed to be active.

Growth Hormone An excess of growth hormone acts on bone by stimulating the growth areas of the skeleton.

Growth hormone excess occasionally occurs, resulting in acromegaly, or giantism. People with acromegaly have large frames, and the bone overgrowth is most apparent in the large hands, feet, and face. Although at first glance bone mass appears to be greater than normal, careful measurements showed that mass was not increased and, in fact, fractures are not uncommon in acromegalic people. Bone overgrowth affecting cartilage at the ends of the bone and the periosteum, (the outside envelope of the bone) was confused with osteosclerosis, or dense bones, and growth hormone was thought of as potentially beneficial for the treatment of osteoporosis. However, studies in osteoporotic individuals showed no increase in bone mass but rather a stimulation of new periosteal bone at the expense of bone from the inner part of the cortex [6].

Agents that Prevent Osteoporosis

Hormones Probably the most extensively studied agents in the treatment of osteoporosis are hormones, both estrogens and androgens. As with growth hormone, phosphates, and the diphosphonates, the basic reasoning behind the use of hormones to treat osteoporosis was wrong. The theory was based on observations that, in egg-laying birds, estrogen fluctuations that occur in the laying cycle control the deposit of calcium stored in the bone marrow. It was concluded that high amounts of estrogens produce new bone deposition, which indeed they do—only in egg-laying birds, however, and not in humans. Estrogens do affect bone, their action being to protect bone against osteoclastic resorption. Therefore, at the time of the menopause and after the menopause, when estrogen levels tend to decline, this protective barrier is lost and bone loss increases. Conversely, in postmenopausal women estrogen will decrease resorption back to "normal" premenopausal levels. Many studies, from those done twenty or thirty years ago to the present day, have repeatedly shown that estrogens given to postmenopausal women will slow bone loss, although over long periods the effect is less.

1,25-Vitamin D The active form of 1,25-dihydroxyvitamin D, the hormone responsible for absorption of calcium in the gut, has also been suggested as being of potential benefit in treating osteoporosis. Over a 6-month period, the effect on bone of 1,25-vitamin D was to decrease bone resorption slightly. Since the study was carried out in Minnesota, it is possible that the osteoporotic patients were slightly deficient in vitamin D and the 1,25-vitamin D corrected the deficiency; in effect, it may have increased calcium absorption and thus depressed resorption. 1,25-Vitamin D does not stimulate bone formation, and therefore there is no increase in bone mass.

Calcium From the discussion of both the importance of a high dietary calcium intake and the relationship between calcium and the secretion of parathyroid hormone, calcium is a logical and scientifically sound agent to use to prevent osteoporosis. An increase in the amount of calcium ingested causes a rise in the serum calcium and a subsequent fall in parathyroid hormone secretion; consequently, bone resorption is suppressed. In osteoporotic patients, calcium supplements, with or without vitamin D, have been found almost to stop bone loss by lowering resorption of bone to a level only a little above that of new bone deposition.

The effect of calcium infusions has also been studied. The direct introduction of calcium into the bloodstream circumvents any inefficiency of absorption in the gut and periodically increases the level of calcium in the serum to high values, thus strongly suppressing parathyroid gland activity. Calcium balance is improved and bone resorption is suppressed. Perhaps surprisingly, the positive balance is maintained after calcium infusions are stopped, suggesting a sustained effect on the parathyroids. However, the use of calcium infusions requires that the patient be hospitalized for the infusions and that the serum calcium be carefully monitored during the infusions. In general, this form of therapy is complex and has little advantage over the use of oral calcium supplements.

Agents that "Cure" Osteoporosis
Osteosclerosis, or dense bone, is found in individuals who live in areas of the world where the natural fluoride level in the soil is high, so that food and water contain high levels of this substance. This observation led early investigators to use fluoride in the treatment of osteoporosis. The results were promising at first; calcium balance improved and large amounts of new bone formed. Long-term studies proved disappointing, however; the positive calcium balance reverted to a negative balance, bone resorption increased, and some investigators reported high levels of parathyroid hormone in the serum. The fluoride obviously stimulated new bone deposition, but this result led to an "abnormal" need for calcium to provide mineral for the new bone. When calcium supplements were not provided, the patients became essentially calcium deficient, and secondary hyperparathyroidism developed.

In animals it was demonstrated that the fluoride-induced osteoid and the stimulation of the parathyroid glands could be prevented by adding calcium to the fluoride. A number of different investigators in a large number of separate studies have consistently shown that a combination of fluoride and calcium will increase new bone formation and decrease bone resorption [7]. The fluoride is responsible for stimulating the osteoblasts, the bone-forming cells that produce two to three times the normal amount of matrix, while the calcium not only is available for deposition in this new matrix so that it becomes mineralized, but also suppresses bone resorption. The net result is an increase in bone mass.

Bone strength studies have clearly shown that the increase in mass is accompanied by an increase in bone strength [8], which will eventually cause a reduction in the number of fractures. Fractures should not be expected to stop immediately when fluoride and calcium are given. Obviously a significant amount of new bone must be formed and bone mass increased to normal before fractures cease. This takes time and depends on the amount of fluoride and calcium given. The combination of 75 mg of sodium fluoride per day, given in divided doses, and 2.4 g of calcium carbonate will produce a bone mass that approaches normal in 2 to 3 years.* Because it is preferable to give the calcium with the fluoride (to be sure fluoride alone is not given and to prevent the gastric discomfort that may occur with fluoride alone) and because calcium decreases the absorption of fluoride to some extent (about 20%), higher doses of fluoride and calcium may be advisable.

CONCLUSIONS

Osteoporosis is a disease that is and will continue to be a problem of increasing importance. It is the result of a response to chronic calcium deficiency, excessive phosphorus, lack of exercise, and many other factors. Since more than one of these factors generally play a role in a single person, the prevention of osteoporosis by one therapeutic agent is unlikely and in fact has not proved to be possible. It is also difficult to convince an asymptomatic person of the need for preventive therapy. Of the agents that slow bone loss and postpone the first fracture and the disease itself, calcium supplements are effective and safe. 1,25-Vitamin D is of dubious efficacy and may cause worrying hypercalcemia. Estrogens are less effective than calcium and have been reported to increase the incidence of endometrial carcinoma, a serious side effect. None of these three agents will "cure" or effec-

---

*Fluoride and calcium may be obtained as Florical from Mericon Industries, 420 S.W. Washington St., Peoria, Illinois 61602.

tively treat a patient who has already lost so much bone that a pathologic fracture has developed. Only fluoride with calcium will achieve this. A fluoride-calcium combination will effectively increase bone mass and thus prevent further fractures. A patient with disability and pain from chronic osteoporosis can become pain-free and lead an active, normal life with this treatment.

REFERENCES

1. Singh, M., Riggs, B.L., Beabout, J.W., Jowsey, J. Femoral trabecular pattern index for evaluation of spinal osteoporosis. *Ann. Intern. Med.* 77:63–67, 1972.
2. Genant, H.K., Gordon, G.S., Hoffman, P.G., Jr. Osteoporosis: Part I. Advanced radiologic assessment using quantitative computed tomography. *West. J. Med.* 139:75–84, 1983.
3. Heaney, R.P., Recker, R.R., Saville, P.D. Calcium balance and calcium requirement in middle-aged women. *Am. J. Clin. Nutr.* 30:1603, 1977.
4. Jowsey, J., Balasubramanian, P. The effect of phosphate supplements on soft tissue calcification and bone turnover. *Clin. Sci.* 42:289–299, 1972.
5. Goldsmith, R.S., Jowsey, J., Dube, W., et al. Effects of phosphorus supplementation on serum parathyroid hormone and bone morphology in osteoporosis. *J. Clin. Endocrinol. Metab.* 43:523–532, 1976.
6. Aloia, J.F., Zanzi, I., Ellis, K., et al. Effects of growth hormone in osteoporosis. *J. Clin. Endocrinol. Metab.* 43:992–999, 1976.
7. Jowsey, J., Riggs, B.L., Kelley, P.J., Hoffman, D.L. Effect of combined therapy with sodium fluoride, vitamin D and calcium in osteoporosis. *Am. J. Med.* 53:43–49, 1972.
8. Franke, J., Runge, H., Grau, P., Fengler, F., et al. Physical properties of fluorosis bone. *Acta Orthop. Scand.* 40:20–27, 1976.

The Kinematics of Aging

▶

R. Donald Hagan, Ph.D.

10

Modern humans possess a wide range of kinematic activities. These activities include numerous modes of hunting, farming, and labor and many forms of recreation, exercise, and sport. The kinematic movements of humans are determined by their physical structure and functional capabilities. These, in turn, govern aerobic and anaerobic energy yield and neuromuscular strength and endurance [1]. It is well-known, however, that as humans age there is a gradual decline in the outward and inward appearance of the body [2]. These changes are associated with a decrease in the integrity of bone, loss of muscle nitrogen and mass, and a decline in nervous, pulmonary, and cardiovascular function, which contribute to a reduction in the muscular strength, cardiopulmonary endurance, and physical performance capabilities of the individual.

In this chapter, changes in the cardiorespiratory factors associated with rest and exercise and alterations in muscular strength with advancing age will be summarized. This discussion will also focus on the effect of physical training on these age-related changes.

CARDIAC FUNCTIONAL CHANGES WITH AGE

Alterations in the structural components of the cardiovascular system with age can greatly affect the functional capacity of the heart, causing the capacity for muscular work to decline [3, 4]. Age-related changes to the heart include changes in heart chamber size, myocardial mass, and geometric dimensions and histologic changes in the valves, conduction apparatus, and large and small coronary vessels. One of the main contributors to this reduction is progressive atherosclerosis of the coronary arteries. Since the heart and systemic vascular system are involved in the delivery of blood and oxygen to active muscle tissue, increases or decreases in physical work capacity will depend on cardiovascular integrity and function.

Many previous investigations examining changes in physical work with age are cross-sectional studies. In this type of investigation it is difficult to eliminate genetic or environmentally induced differences between groups. Unfortunately, few longitudinal studies have been conducted concerning changes in physical work capacity with advancing age. Furthermore, the effect of exercise conditioning and its influence on the decline in work capacity with age is not well understood.

Numerous cross-sectional studies on sedentary men and women indicate that the decline in maximal aerobic power, as indicated by the maximal oxygen uptake capacity ($\dot{V}O_2$max), is relatively constant after age 20 years. The concept of constancy in the reduction of $\dot{V}O_2$max with age is based on best-fit linear regression equations. Hodgson and Buskirk [5] reviewed 23 studies involving 1800 observations of men aged 18 to 60 years and reported a decline in $\dot{V}O_2$max that averaged 0.41 ml/kg/min/yr. In addition, Shephard [6] reported a decrease of 0.45 ml/kg/min/yr, while Dehn and Bruce [7] reported that the reduction calculated from 17 studies involving approximately 700 observations was 0.4 ml/kg/min/yr. In sedentary women, Profant et al. [8] and Drinkwater et al. [9] reported a decrease of 0.3 ml/kg/min/yr.

Similar studies conducted in moderately active and active athletic men indicate that $\dot{V}O_2$max also declines with age in these groups. For moderately active men, the decline is 0.44 ml/kg/min/yr [5], and for marathon runners, a value of 0.27 ml/kg/min/yr has been reported [10, 11]. In a similar study, conducted on 67 male marathoners between the ages of 20 and 61 years, I found a decrease of 0.4 ml/kg/min/yr (Fig. 10-1).

Caution is advised when using regression equations to predict the entire effects of age on maximal aerobic power. It is important to emphasize that these equations may convey an oversimplified view of the modifying influence of physical activity on cardiorespiratory functional capacity. In another cross-sectional study, Matter et al. [12] reported that $\dot{V}O_2$max was relatively constant until age 45 years, after which the decline averaged 0.5 ml/kg/min/yr. Other studies indicate a more varied rate of decline for maximal aerobic power with age. Dehn and Bruce [7] followed 40 sedentary men for 2.5 years and reported that the decline in $\dot{V}O_2$max was 0.94 ml/kg/min/yr. Dehn and Bruce also viewed the

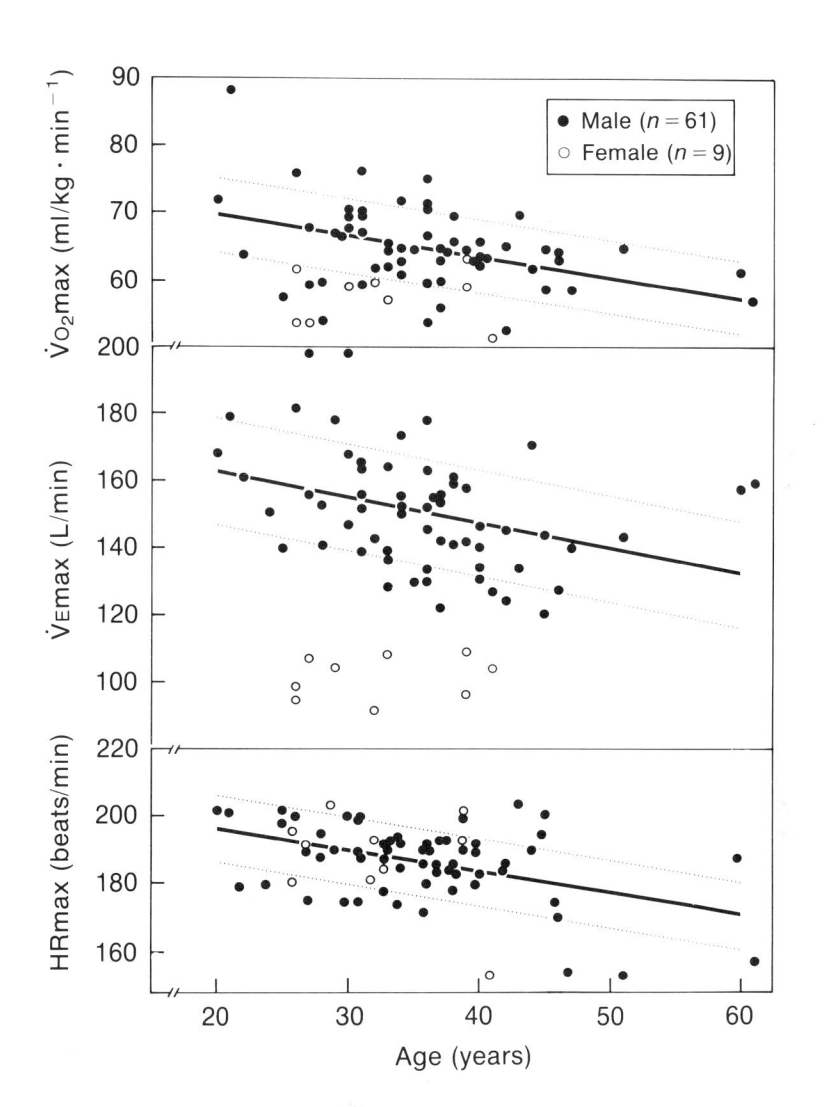

Figure 10-1

Maximal oxygen uptake, minute ventilation, and heart rate in relation to age in male and female marathon runners. Solid circles represent men ($n = 61$), open circles represent women ($n = 9$). For the man: $\dot{V}O_2$max, ml/kg/min = 78.6 − 0.4 (age in years); SEE = ±5.0, $r = -0.51$, $p < 0.0001$; HRmax, beats/min = 207.6 − 0.6 (age in years); SEE = ±12.0, $r = -0.40$, $p < 0.0001$.

work of Hollman and associates, who found the decrease to be 0.93 ml/kg/min/yr in 56 sedentary men followed for over 12 to 15 years. Irving et al. [13] measured the maximal aerobic power on three separate occasions over 8 years in 12 healthy men. The decline in $\dot{V}O_2$max was 0.96 ml/kg/min/yr for sedentary men and 0.79 ml/kg/min/yr for the physically active men. Dill et al. [14] followed 13 former champion runners and found a rate of decline of 1.04 ml/kg/min/yr. Interestingly, when 9 of these men were studied 8 to 9 years later, the reduction was 0.52 ml/kg/min/yr [15, 16].

The decline in maximal aerobic power with age has been explained by a decline in cardiovascular reserve. Numerous cross-sectional studies indicate that, with advancing age, maximal exercise heart rate declines [4, 17–19]. Most of these studies indicate the rate of decrease is from 0.5 to 1.0 beat/yr. Increased fat, elastic, and fibrous tissue in the sinoatrial and atrioventricular conduction nodes and their bundle branches indicate that the age-related decline in maximal exercise heart rate is associated with changes in the intrinsic structure of the heart [20]. In addition, the maximal stroke volume and ejection fraction and maximal cardiac output decline with age [20–24], and the maximal systemic arteriovenous oxygen difference is lower in the aged [20, 25].

AEROBIC TRAINING WITH AGE

As an individual advances in age, he usually becomes more sedentary [26]. Thus, one of the important issues concerning the age-related decline in maximal aerobic power is the effect of exercise conditioning. It is well-known that exercise conditioning, especially aerobic exercise such as walking, running, or bicycling, produces an increased maximal exercise stroke volume, maximal cardiac output, and maximal arteriovenous oxygen difference, leading to an increase in maximal oxygen uptake and physical work capacity [26, 27]. The chronologic age of an individual does not appear to be a deterrent to exercise conditioning, especially endurance training [23, 28–31]. The $\dot{V}O_2$max of older men engaged in endurance training is significantly greater than that of sedentary men of similar age [23, 28–30, 32–34]. Recent studies indicate that the relative change in $\dot{V}O_2$max with training in older individuals is similar to that in younger age groups [12, 35, 36], with increases in stroke volume, cardiac output, and arteriovenous oxygen difference as the mechanism behind the increased work capacity. However, the older individual may need a longer period to adapt to training [35]. Although $\dot{V}O_2$max decreases with age, increases in percentage of body fat and total body weight [37-39] may force individuals to reduce their training loads. More investigation is needed to evaluate the effects of long-term training on changes in cardiorespiratory functional capacity with age before definite statements can be formulated [35].

MUSCULOSKELETAL AND NEUROMUSCULAR CHANGES WITH AGE

Changes in the skeletal structure, muscle content, and nervous system will collectively produce reductions in strength and endurance with age.

A person's outward appearance can be a good indicator of the rate of aging of their musculoskeletal system [40]. The posture becomes stooped, with the head and neck held forward. This has been attributed to muscle shrinkage, decreases in elasticity, and calcification of ligaments and tendons. The intervertebral disks flatten, leading to a reduction in height. In general, there is progressive negative calcium balance with advancing age; however, much of this can be attributed to the disease of osteoporosis.

The muscles decline in number of muscle fibers, size, and mass, with fibrous tissue secondarily replacing the contractile units. The reduced number of myofibrils is associated with an increased negative nitrogen balance, which leads to a decrease in muscular strength and endurance. In the nervous system, there is an age-related loss in the total number of brain cells and their fibers, as well as an increase in the rigidity of remaining brain tissues. It has been reported that a 20 to 25 percent loss in brain weight occurs between the ages of 20 and 90 years [32]. Conduction velocity of nerve impulses also decreases with age, leading to slower reactions and voluntary motor movement [41].

Muscular strength and endurance are related primarily to total lean muscle mass, cross-sectional size of the muscle, the number of active motor units recruited and the rate of fiber contractions, neuromuscular coordination, and blood supply. All of these factors have been shown to be affected by the age of the individual. Part of the problem, however, stems from the difficulty in separating specific age-related changes in muscle from those from physical activity patterns, disease, and cardiovascular, hormonal, and neural factors.

Skeletal muscle ultrastructure and electrophysiology are primarily determined by the type of neural innervation. The type of innervation produces three basic types of muscle fibers. These types include fast-twitch high-oxidative fibers, fast-twitch low-oxidative fibers, and slow-twitch high-oxidative fibers. With aging, fast-twitch fibers are characterized by a decrease in both number and size, while slow-twitch fibers decrease in number [36]. In addition, the ratio of glycolytic to oxidative enzymes in the fast-twitch fibers decreases [36]. This could lead to a reduced capacity for anaerobic energy yield.

The physiologic changes associated with the age-related decline in muscular strength suggest that the neuromuscular junction is involved. Gutman et al. [42] reported no evidence of disintegration of the terminal axons in the senile muscle fibers of rats but did observe major changes in the neuromuscular junction. These changes included an increase in the number and agglutination of presynaptic vesicles, appearance of neurotubules and neurofilament in the peripheral axons, enlargement of primary synaptic clefts, thickening of the junctional folds, and increased branching of the junctional folds. Also, there was a reduced frequency of miniature end-plate potentials and reduced conduction velocity in the presynaptic axon. Furthermore, aging was associated with decreases in muscle number, diameter, and mass, with the proximal muscle of the lower limb particularly affected.

MUSCULAR STRENGTH CHANGES WITH AGE

Muscular strength is defined as the greatest amount of force that skeletal muscle can produce in a single maximal concentric contraction, while *strength endurance* refers to the maximal number of lifting repetitions that can be produced with a known resistance. Strength and endurance can be measured as isometric force in which no lever movement occurs, as isotonic force in which lever movement and muscle tension occur at variable rates, and as isokinetic force in which lever movement is set at a constant, but muscle tension varies.

The great majority of studies evaluating changes in muscle strength and endurance with age are cross-sectional studies. These reports indicate that dominant and subdominant handgrip isometric strength and 1-minute sustained isometric contractions reach their peak force between 20 and 30 years of age and then decline progressively (approximately 0.2 kg/yr) in both men and women [4, 43], so that by age 65 the strength values are approximately 80 percent of that attained between 20 and 30 years of age [4]. It has also been reported that the grip strength in men is reduced at a greater rate between age 60 and 80 than between 30 and 60 years of age [5]. A significant and linear decrease in isometric grip strength with age at a rate of 0.23 kg/yr has been reported for healthy women, aged 19 to 65 years [5]. How-

The Kinematics of Aging

ever, it has also been reported that there is an increase in the duration of a sustained 40 percent of maximum isometric contraction.

In addition, Larsson et al. [44] have evaluated the isometric and isokinetic muscle strength and speed of movement in relation to muscle morphology and age in sedentary males between 11 and 70 years of age. Isometric and dynamic isokinetic strength increased up to the third decade, remained constant to the fifth decade, and then decreased with increasing age. There was no measurable external atrophy of the quadriceps muscle, but histochemical changes indicated a decreased proportion of and a selective atrophy of fast-twitch fibers with increasing age.

Cross-sectional studies that I have conducted have involved the measurement of isokinetic knee extension and flexion, isokinetic bench press and leg press strength, and isotonic bench press and leg press strength and determination of low back and hamstring flexibility (Tables 10-1 through 10-4). The findings indicate that, in adults, isokinetic

Table 10-1
Isokinetic Right and Left Knee Extension and Flexion, and Quadriceps-to-Hamstring Ratio Adjusted for Body Mass Index by Decade in Men Aged 20 to 69 Years[a]

Age (years)	n	Right Knee Extension (kg)	Left Knee Extension (kgm)	Right Knee Flexion (kgm)	Left Knee Flexion (kgm)	Right Quad-Ham (%)	Left Quad-Ham (%)
20–29	37	20.7 ± 3.4	20.6 ± 3.4	11.3 ± 2.5	11.1 ± 2.5	65 ± 1[b]	65 ± 1[b]
30–39	154	19.8 ± 3.4	19.2 ± 3.4	11.5 ± 3.4	11.1 ± 3.4	63 ± 1	64 ± 1
40–49	173	18.2 ± 3.6[c]	17.7 ± 3.6[c]	10.4 ± 3.6[c]	10.0 ± 3.6[c]	64 ± 1	64 ± 1
50–59	101	16.2 ± 4.2[c]	15.5 ± 4.2[c]	9.3 ± 2.8	8.8 ± 2.8	64 ± 1	64 ± 1
60–69	46	14.4 ± 3.7[c]	13.8 ± 3.7[c]	8.8 ± 2.8	8.7 ± 2.8	62 ± 1	62 ± 1

Body mass index = wt/ht^2; quad-ham ratio = quad peak force/quad + ham peak force; kgm = kilogram-meters.
[a]Values represent means ± S.D. at 60°/sec.
[b]$p < 0.05$ compared to 60–69 age group.
[c]$p < 0.05$ compared to all age groups.

and isotonic strength reaches peak values in the 20s and declines every decade thereafter.

Isokinetic knee extension strength decline in both men and women averaged approximately 0.155 kg/yr, while the decline in knee flexion strength averaged about 0.07 kg/yr. Interestingly, however, the quadriceps-to-hamstring ratios in both males and females are maintained at 64 to 65 percent through the seventh decade.

Bench press strength measured both isotonically and isokinetically declined in males at a rate of 0.82 kg/yr and 0.2 kg/yr, respectively and in females at a rate of 0.36 kg/yr and 0.1 kg/yr, respectively. Leg press strength measured both isotonically and isokinetically declined in males at a rate of

Table 10-2
Isotonic and Isokinetic Bench and Leg Press Strength and Low Back–Hamstring Flexibility Adjusted for Body Mass Index by Decade in Men Aged 20 to 69 Years[a]

Age (years)	n	Isotonic Bench Press (kg)	Isokinetic[b] Bench Press (kgm)	Isotonic Leg Press (kg)	Isokinetic[b] Leg Press (kgm)	Low Back–Hamstrings Sit-Reach (cm)
20–29	37	86.4 ± 16.6[c]	23.8 ± 5.0	151.8 ± 19.4[c]	73.0 ± 11.8	44.2 ± 10.8
30–39	154	77.7 ± 16.9[c]	23.0 ± 5.1	137.3 ± 22.6[c]	70.5 ± 12.0	42.4 ± 9.5
40–49	173	67.7 ± 17.9[c]	20.2 ± 5.4[c]	131.8 ± 17.9[c]	65.0 ± 10.9[c]	37.8 ± 10.0[c]
50–59	101	59.5 ± 13.7[c]	17.8 ± 5.6[c]	120.0 ± 18.2	58.3 ± 11.1[c]	37.1 ± 10.2
60–69	46	54.5 ± 15.4[c]	15.8 ± 5.6[c]	115.4 ± 18.5	53.9 ± 11.2[c]	35.6 = 10.3

Body mass index = wt/ht^2; kgm = kilogram-meters.
[a]Values represent means ± S.D.
[b]Isokinetic measures conducted at 60°/sec.
[c]$p < 0.05$ compared to all age groups.

Table 10-3
Isokinetic Right and Left Knee Extension and Flexion, and Quadriceps-to-Hamstring Ratio Adjusted for Body Mass Index by Decade in Women Aged 20 to 59 Years[a]

Age (years)	n	Right Knee Extension (kgm)	Left Knee Extension (kgm)	Right Knee Flexion (kgm)	Left Knee Flexion (kgm)	Right Quad-Ham (%)	Left Quad-Ham (%)
20–29	23	12.0 ± 2.0[b]	11.9 ± 2.0[b]	6.4 ± 1.3	6.4 ± 1.3	66 ± 1	65 ± 1
30–39	35	10.2 ± 1.4[c]	9.8 ± 1.4[c]	5.5 ± 1.6[c]	5.2 ± 1.6[c]	65 ± 1	65 ± 1
40–49	16	10.2 ± 2.2	9.0 ± 2.2	5.5 ± 1.6	5.2 ± 1.6	64 ± 1	64 ± 1
50–59	11	7.3 ± 2.3	7.1 ± 2.3	4.0 ± 1.4[b]	3.7 ± 1.4[b]	65 ± 1	65 ± 1

Body mass index = wt/ht^2; quad-ham ratio = quad peak force/quad + ham peak force; kgm = kilogram-meters.
[a]Values represent means ± S.D. at 60°/sec.
[b]$p < 0.05$ compared to all age groups.
[c]$p < 0.05$ compared to 20–29 and 50–59 age groups.

0.9 kg/yr and 0.47 kgm/yr, respectively and in females at a rate of 0.14 kg/yr and 0.68 kgm/yr. Flexibility of the low back and hamstrings is better maintained in females with the decline approximately 0.15 cm/yr compared to 0.2 cm/yr for males.

Table 10-4
Isotonic and Isokinetic Bench and Leg Press Strength and Low Back–Hamstring Flexibility Adjusted for Body Mass Index by Decade in Women Aged 20 to 59 Years[a]

Age (years)	n	Isotonic Bench Press (kg)	Isokinetic[b] Bench Press (kgm)	Isotonic Leg Press (kg)	Isokinetic[b] Leg Press (kgm)	Low Back–Hamstrings Sit-Reach (cm)
20–29	23	38.2 ± 6.5[c]	8.2 ± 2.6[c]	83.2 ± 13.0[c]	46.7 ± 9.9[c]	50.5 ± 7.3
30–39	35	34.1 ± 8.1[d]	6.6 ± 3.3	75.9 ± 13.4[d]	41.1 ± 9.8[d]	48.8 ± 7.5
40–49	16	30.9 ± 7.3	5.4 ± 2.8	70.9 ± 12.7	39.1 ± 10.0	46.7 ± 7.1
50–59	11	27.3 ± 7.5	5.1 ± 2.8	52.7 ± 12.1	26.4 ± 9.6	46.0 ± 6.7

Body mass index = wt/ht^2; kgm = kilogram-meters.
[a]Values represent means ± S.D.
[b]Isokinetic measures conducted at 60°/sec.
[c]$p < 0.05$ compared to all age groups.
[d]$p < 0.05$ compared to 20–29 and 50–59 age groups.

STRENGTH TRAINING WITH AGE

While cross-sectional studies indicate a gradual decrease in muscular strength with age, the effects of muscular strength training on muscle size and function are not well-known. Surely a gradual sedentary life-style must play a part in the role of reduction of muscle strength of older individuals. Histochemical and biochemical studies in humans indicate that immobilization of a limb can promote loss in fiber diameter, reduced high-energy phosphate concentrations, and reduced glycogen content. Thus, inactivity and disuse could produce a reduction in functional capacity similar to the progressive changes observed in the neuromuscular system of the aged. However, recent studies suggest that strength training can significantly alter the muscle fiber status of the older individual. Larsson [45] evaluated the influence of physical training on muscle morphology and strength in men 22 to 65 years of age. Age-related muscle fiber atrophy, seen before training, was diminished after training because of an increase in fiber size in the older individuals. However, muscular strength values were lower with increasing age before and after training, suggesting that strength decline with increasing age is not due to muscle fiber atrophy.

Neural factors as opposed to muscle hypertrophy have been proposed as the dominant factor in strength gain. Moritani [36] reported that significant increases in the maximal strength in the aged were brought about by an increase in the maximal muscle activation level without changes in cross-sectional size of the muscle. Thus, strength training in older individuals may be due to neural factors, producing a greater level of muscle activation rather than hypertrophy.

CONCLUSIONS

Maximal aerobic power and muscular strength have been shown in many cross-sectional studies to decline with advancing age. However, there are many examples of high-level physical strength and endurance in older individuals. The level of daily physical activity seems to be a very important factor in human functional capacity. While physical exercise cannot reverse the aging process, it can increase functional capacity. This effect on the aging individual needs to be studied further.

REFERENCES

1. Astrand, P.-O. Physical performance as a function of age. *J.A.M.A.* 205:105–109, 1968.
2. Smith, E.L. Age: the interaction of nature and nurture. In Smith, E.L., Serfass, R.C. (eds.), *Exercise and Aging: The Scientific Basis*. Hillside, N.J.: Enslow Publishers, 1980, pp. 11–17.
3. Astrand, I., Astrand, P.-O., Hallback, I., Kilbom, A. Reduction in maximal oxygen uptake with age. *J. Appl. Physiol.* 35:649–654, 1973.
4. Astrand, P.-O., Rodahl, K. *Textbook of Work Physiology*. New York: McGraw-Hill, 1977, p. 122.
5. Hodgson, J.L., Buskirk, E.R. Physical fitness and age, with emphasis on cardiovascular function in the elderly. *J. Am. Geriatr. Soc.* 15:385–392, 1977.
6. Shephard, R.J. Cardiovascular limitations in the aged. In Smith, E.L., Serfass, R.C. (eds.), *Exercise and Aging: The Scientific Basis*. Hillside, N.J.: Enslow Publishers, 1980, pp. 19–29.
7. Dehn, M.M., Bruce, R.A. Longitudinal variations in maximal oxygen intake with age and activity. *J. Appl. Physiol.* 33:805–807, 1972.
8. Profant, G.R., Early, R.G., Nilson, K.L., et al. Responses to maximal exercise in healthy middle-aged women. *J. Appl. Physiol.* 33:595–599, 1972.
9. Drinkwater, B.L., Horvath, S.M., Wells, C.L. Aerobic power of females, ages 10 to 68. *J. Gerontol.* 30:385–394, 1975.
10. Cooper, K.H., Purdy, J.L., White, S.R., et al. Age-fitness adjusted maximal heart rates. *Med. Sport* 10:78–88, 1977.
11. Costill, D.L., Winrow, E. Maximal oxygen intake among marathon runners. *Arch. Phys. Med. Rehabil.* 51:317–320, 1970.

12. Matter, S., Stamford, B.A., Weltman, A. Age, diet, maximal aerobic capacity and serum lipids. *J. Gerontol.* 35:532–536, 1980.
13. Irving, J.B., Kusumi, F., Bruce, R.A. Longitudinal variations in maximal oxygen consumption in healthy men. *Clin. Cardiol.* 3:134–136, 1980.
14. Dill, D.B., Robinson, S., Ross, J.C. A longitudinal study of 16 champion runners. *J. Sports Med.* 7:4, 1967.
15. Robinson, S., Dill, D.B., Robinson, R.D., et al. Physiological aging of champion runners. *J. Appl. Physiol.* 41:46–51, 1976.
16. Robinson, S., Dill, D.B., Ross, J.C., et al. Training and physiological aging in man. *Fed. Proc.* 32:1628–1634, 1973.
17. Astrand, I., Astrand, P.-O., Rodahl, K. Maximal heart rate during work in older men. *J. Appl. Physiol.* 14:562–566, 1959.
18. Robinson, S. Experimental studies of physical fitness in relation to age. *Arbeitsphysiologie* 10:251–323, 1938.
19. Sheffield, L.T., Maloof, J.A., Sawyer, J.A., Roitman, D. Maximal heart rate and treadmill performance of healthy women in relation to age. *Circulation* 57:79–84, 1978.
20. Gerstenblith, G., Lakatta, E.G., Weisfeldt, M.L. Age changes in myocardial function and exercise response. *Prog. Cardiovasc. Dis.* 19:1–21 1976.
21. Hossack, K.F., Bruce, R.A., Green, B., et al. Maximal cardiac output during upright exercise: approximate normal standards and variations with coronary heart disease. *Am. J. Cardiol.* 46:204–212, 1980.
22. Hossack, K.F., Kusumi, F., Bruce, R.A. Approximate normal standards of maximal cardiac output during upright exercise in women. *Am. J. Cardiol.* 47:1080–1086, 1981.
23. Pollock, M.L., Miller, H.S., Jr., Wilmore, J. Physiological characteristics of champion American track athletes 40 to 75 years of age. *J. Gerontol.* 29:645–649, 1974.
24. Port, S., Cobb, F.R., Coleman, R.E., Jones, R.H. Effect of age on the response of the left ventricular ejection fraction to exercise. *N. Engl. J. Med.* 303:1133–1137, 1980.
25. Kanstrup, I., Ekblom, B. Influence of age and physical activity on central hemodynamics and lung function in active adults. *J. Appl. Physiol.* 45:709–717, 1978.
26. Niinimaa, V., Shephard, R.J. Training and oxygen conductance in the elderly: I. The respiratory system. *J. Gerontol.* 33:354–361, 1978.
27. Saltin, B., Blomquist, G., Mitchell, J.M., et al. Response to exercise after bed rest and after training. American Heart Association Monograph No. 23, 1968.
28. Dill, D.B. Marathoner DeMar: physiological studies. *J. Natl. Can. Inst.* 35:185–190, 1965.
29. Pollock, M.L., Miller, H.S., Jr., Linnerud, A.C., et al. Physiological findings in well-trained middle-aged American men. *Br. Assoc. Sports Med. J.* 7:222–228, 1973.
30. Pollock, M.L., Miller, H.S., Ribisl, P.M. Effect of fitness on aging. *Physician Sportsmed.* 8:45–48, 1978.
31. Schmidt, M.N., Wrenn, J.P. Selected physical and cardiorespiratory parameters of active males, aged 40–59. *J. Sports Med. Phys. Fitness* 18:183–188, 1978.
32. Rockstein, M. The biology of aging in humans—an overview. In Goldman, R., Rockstein, M. (eds.), *The Physiology and Pathology of Human Aging.* New York: Academic Press, 1975, pp. 1–7.
33. Rockstein, M., Chesky, J.A., Lopez, T. Effects of exercise on the biomedical aging of mammalian myocardium: I. Actomyosin ATPhase. *J. Gerontol.* 36:294–297, 1981.

34. Wilmore, J.H., Miller, H.L., Pollock, M.L. Body composition and physiological characteristics of active endurance athletes in their eighth decade of life. *Med. Sci. Sports* 29:44–48, 1974.
35. American College of Sports Medicine. The recommended quantity and quality of exercise for developing and maintaining fitness in healthy adults. Position statement, Madison, Wisc., 1981.
36. Moritani, T. Training adaptations in the muscles of older men. In Smith, E.L., Serfass, R.C. (eds.), *Exercise and Aging: The Scientific Basis*. Hillside, N.J.: Enslow Publishers, 1980, pp. 149–166.
37. Parizkova, J., Eiselt, E. Longitudinal changes in body build and skinfolds in a group of men over a 16-year period. *Hum. Biol.* 52:803–809, 1980.
38. Tzankoff, S.P., Norris, A.H. Effect of muscle mass decrease on age-related BMR changes. *J. Appl. Physiol.* 43:1001–1006, 1977.
39. Tzankoff, S.P., Norris, A.H. Longitudinal changes in basal metabolism in man. *J. Appl. Physiol.* 45:536–539, 1978.
40. Fitts, R.H. Aging and skeletal muscle. In Smith, E.L., Serfass, R.C. (eds.), *Exercise and Aging: The Scientific Basis*. Hillside, N.J.: Enslow Publishers, 1980, pp. 31–44.
41. Spirduso, W.W. Physical fitness, aging, and psychomotor speed: a review. *J. Gerontol.* 35:850–865, 1980.
42. Gutman, E., Hanzilikova, V., Vyskocil, F. Age changes in cross-striated muscle of the rat. *J. Physiol. (Lond.)* 216:331–343, 1971.
43. Norris, A.H., Shock, N.W. Exercise in the adult years—with special reference to the advanced years. In Johnson, W.R. (ed.), *Science and Medicine of Exercise and Sports*. New York: Harper and Brothers Publishers, 1960, pp. 466–490.
44. Larsson, L., Grimby, G., Karlsson, J. Muscle strength and speed of movement in relation to age and muscle morphology. *J. Appl. Physiol.* 46:451–456, 1979.
45. Larsson, L. Physical training effects on muscle morphology in sedentary males at different ages. *Med. Sci. Sports Exercise* 14:203–206, 1982.

The Aging Lumbar Spine

▶

Dan M. Spengler, M.D.

11

EPIDEMIOLOGY

Pain complaints that relate to the spine are the most common symptoms affecting people of industrialized countries. Men and women seem to be affected equally [1]. In both sexes, however, the trend is toward a decrease in pain complaints with increasing age [1] (Fig. 11-1). The decrease in symptoms occurs in spite of the progressive adverse changes in the composition and mechanics of the components of the spine. The lack of any clear-cut relationship between subjective pain complaints and measurable changes in the biochemistry or biomechanics of the components of the spine continues to challenge investigators. In addition, differentiation of degenerative changes from the "normal" consequences of aging is not possible.

Age-related changes in the spine do result in a decrease in the amount of motion present and alter the overall alignment of the spine [1] (Table 11-1). Neurologic abnormalities that are secondary to degenerative changes in the spine also increase with age in both men and women [1]. A significant increase in the percentage of restriction of motion in men as compared to women in anteflexion and retroflexion and rotation is shown in Table 11-1. In addition, a decrease in lordosis of the lumbosacral spine is more common in men than women in the 75-years-plus age group [1].

COMPONENTS OF THE SPINE

Intervertebral Disk

The intervertebral disk is composed of the nucleus pulposus, annulus fibrosus, and the superior and inferior cartilaginous end-plates. Changes in gross morphology of the intervertebral disk with aging can be visualized by examining Figures 11-2 and 11-3. These two illustrations are cross-sections of the intervertebral disk at the L4 level. The dramatic changes in the structure of the disk with age are well demonstrated.

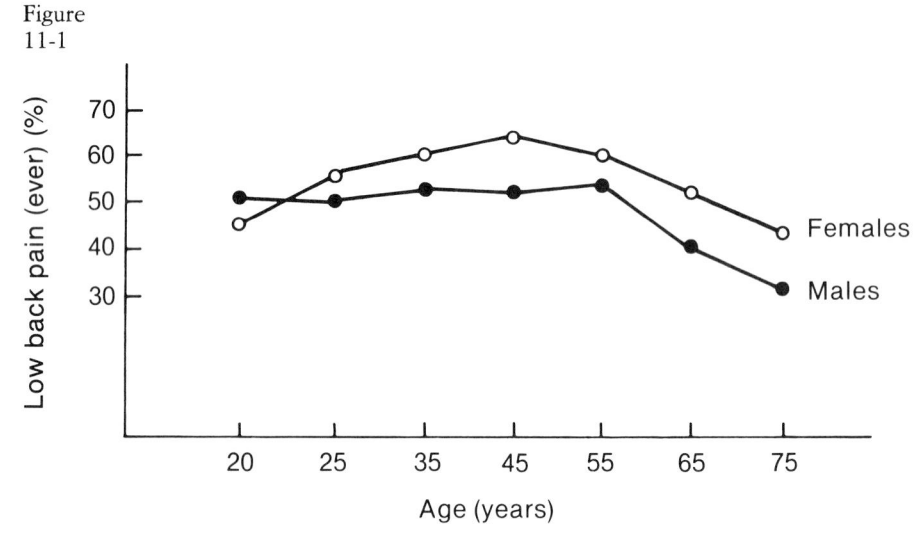

Figure 11-1

The incidence of low back pain plotted against the age of the patient population in years. Peak incidence of low back pain in both sexes appears to occur in the age range from 35 to 55 years. Note the diminishing incidence of low back pain with increased age.
SOURCE: Reproduced with permission from Valkenburg and Haanen [1].

Table 11-1
Age-Specific Physical Findings in Men and Women

Age (years)	Restriction (%) Ante-/Retroflexion	Restriction (%) Rotation	Thoracic Kyphosis (%)	Lumbar Lordosis ↑ (%)	Lumbar Lordosis ↓ (%)	Neurologic Abnormality (%)
20–24	3.1 (5.4)	0.3 (2.0)	1.4 (2.0)	3.1 (4.7)	1.0 (1.0)	0 (1.0)
45–54	18.7 (9.8)	9.7 (9.4)	5.6 (6.7)	2.8 (2.5)	4.5 (3.1)	2.1 (1.5)
75+	64.0 (46.9)	59.3 (43.3)	44.2 (45.9)	5.8 (17.0)	16.3 (6.7)	3.5 (1.6)

Three representative age groups are shown. Note the increased restriction in both anteflexion/retroflexion and rotation with increasing age. The percentage of persons exhibiting lumbar lordosis is observed in both sexes, but more significantly in men. The percentage of persons exhibiting neurologic abnormality increases with age but, again, is more likely in men.
[a]Data for women are in parentheses.
SOURCE: Reproduced with permission from Vaalkenburg and Haanen [1].

The number and size of collagen fibers in the intervertebral disk increase with age [2]. The distinguishing features of the nucleus pulposus disappear as age transfers the entire disk into fibrocartilage [2]. In addition, the aging nucleus shrinks in volume and displaces posteriorly. Biochemical changes that occur in the aging disk include a decrease in water con-

Figure 11-2

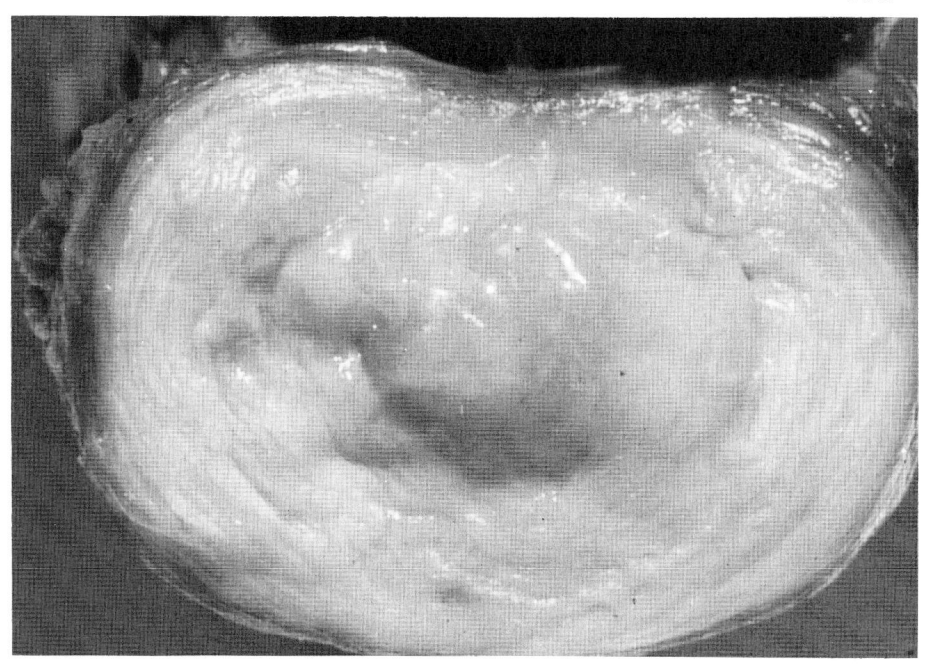

A cross-section through the intervertebral disk of a 17-year-old person. Note the obvious demarcation between the nucleus pulposus and the surrounding annulus fibrosus. The gelatinous consistency of the nucleus can also be identified.
SOURCE: Reproduced with permission from Kirkaldy-Willis et al. [17].

tent, in keratin sulfate, and in the ratio of chondroitin-4 to chondroitin-6 sulfate [3, 4]. Glycoproteins, however, increase in the intervertebral disk with age [2] (Table 11-2).

Nachemson, Schultz, and Berkson [5] demonstrated that differences in mechanical behaviors due to differences in age are not pronounced (Fig. 11-4). Galante [6] examined the tensile properties of the annulus fibrosus in relation to both aging and degenerative changes. Galante observed that after the third decade the tensile properties of the annulus were not influenced by age. Degeneration, however, did induce significant changes in the load-deformation responses of the tissue, decreasing its mechanical efficiency. More recently, Maroudas and coworkers [7] have emphasized the importance of the integrity of the cartilage end-plate cells for the proper nourishment of the nucleus pulposus.

Facet Joint
The facet joints of the spine, which are true synovial joints, have been implicated as possible pain generators [8]. In an

Table 11-2

Biochemical Changes in the Aging Disk

Water content $\downarrow$

Keratin SO_4 $\downarrow$

Ratio of $\dfrac{\text{chondroitin-4-SO}_4}{\text{chondroitin-6-SO}_4}$ $\downarrow$

Glycoproteins $\uparrow$

Figure 11-3

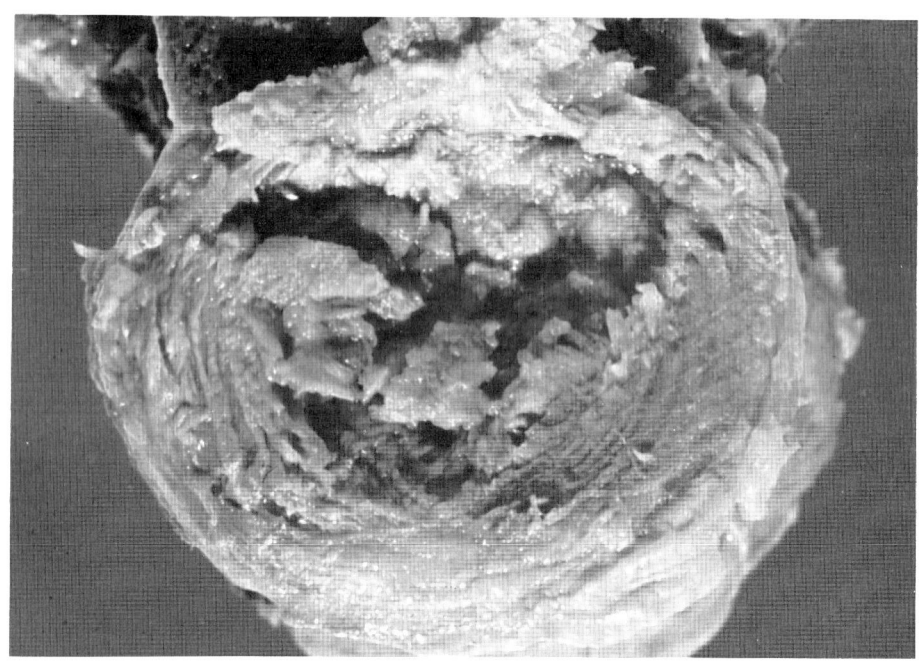

Cross-section of an intervertebral disk of the lumbar spine of a patient in the eighth decade. Note the lack of demarcation between the nucleus pulposus and the annulus fibrosus. Note also the significant radial tears throughout the annulus. Compare this figure with Figure 11-2.
SOURCE: Reproduced with permission from Kirkaldy-Willis et al. [17].

experimental model, Sullivan and Farfan [9] demonstrated that removal of the facets will lead to an acceleration of disk degeneration in rabbits. The authors theorized that this alteration in the intervertebral disk was due to increased torsion loads in the disk. Nachemson [10] noted that the lumbar

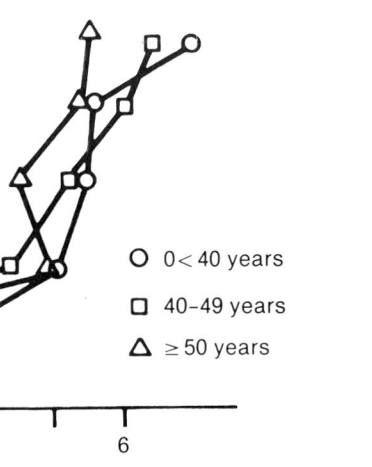

Figure 11-4

Angular deformation curve for intervertebral disk units from individuals of various ages. Note the similarities of the curves.
SOURCE: Reproduced with permission from Nachemson, Schultz, and Berkson [5].

facet joints accounted for approximately 18 percent of load sharing in compression. Farfan [11] demonstrated that the facet joints share 50 percent of the applied load in shear and approximately 50 percent of the load in rotation. Load sharing in the facet joints is diminished with advancing degenerative changes.

Bone

Cancellous and cortical bone play important roles in the transmission and distribution of stresses in the human skeleton, particularly in the vertebral column. Carter and Hayes [12] have considered all bone (cancellous and cortical) as a single material whose apparent density varies over a wide range. These authors demonstrated that the compressive strength of bone is proportional to the square of the apparent density. The compressive modulus is proportional to the cube of the apparent density (Fig. 11-5). The most striking feature of these data is that they apply to the entire range of apparent density in the skeleton, from cortical bone to the most porous cancellous bone [12].

Burstein, Reilly, and Martens [13] examined the age-related changes of human cortical bone. These authors reported a 5- to 7-percent decrease in plastic strain from the third to the ninth decades. They speculated that the observed

Figure 11-5

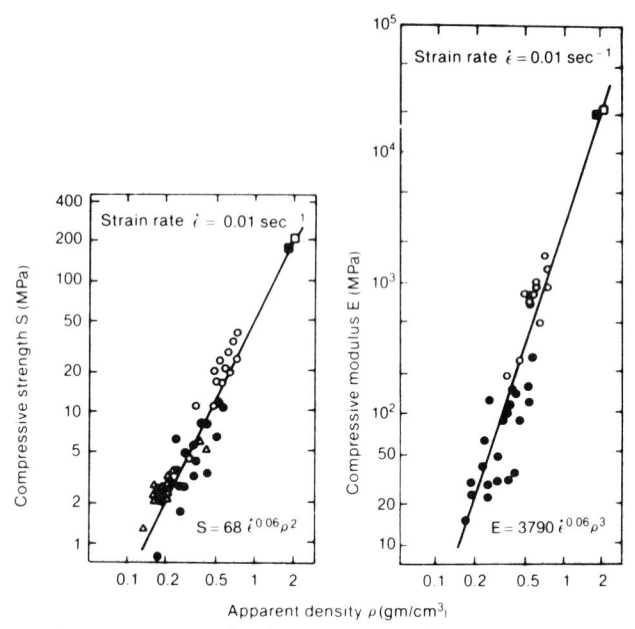

Influence of bone apparent density on compressive strength and modulus.
SOURCE: Reproduced with permission from Carter and Hayes [12].

Figure 11-6

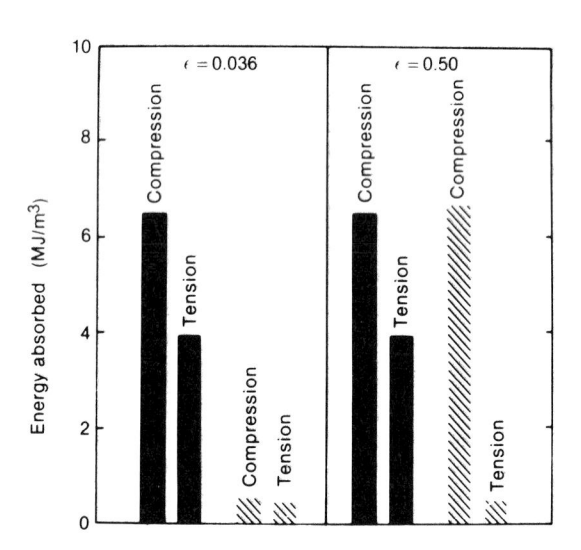

The energy absorption capacity of cortical (solid bars) and cancellous (hatched bars) bone at low strains (strain = 0.036) and at high strains (strain = 0.5). The assumed apparent density of the cancellous bone is 0.4 g/cc.
SOURCE: Reproduced with permission from Carter et al. [14].

decrease in plastic strain may have been related to an increase in collagen cross-linking of bone matrix.

The importance of trabecular bone in energy absorption was highlighted by the study of Carter, Schwab, and Spengler [14]. These authors demonstrated that at a strain rate of approximately 50 percent (as may be encountered with severe trauma) the energy absorbed in fracture of cancellous bone in compression can exceed that absorbed in fracture of cortical bone under any loading condition (Fig. 11-6). The ability of the spine to transmit load will therefore be decreased by the trend toward decreased bone density with increasing age.

Other Components
Other important components of the spine complex include muscles, tendons, and ligaments. These structures are all important in the transmission of forces across the spine and intervertebral unit. Likewise, these tissues are subject to changes with aging. Further work is necessary to delineate the specific role of these structures in aging alterations of the spine. In addition, the nervous system, including the spinal cord, cauda equina, and peripheral nerve roots, is housed within the vertebral canal. Thus, the weakening of the osseous and ligamentous structures with increasing age subjects the older person to an increased risk of injury to the nervous system.

PATHOLOGIC CONSEQUENCES OF AGING

Osteoporosis
Osteoporosis, a decrease in bone mass with age, results in an increased likelihood of bone failure from externally applied forces. Indeed, when the spine becomes severely osteoporotic, fractures of the vertebral bodies may occur spontaneously with minimal or no trauma (Fig. 11-7). Preventing the loss of bone mass with age remains an exciting challenge for investigators.

Disk Degeneration
While most individuals exhibit age-related changes of the spine, only a small percentage develop symptoms that persist and necessitate surgical considerations. Nevertheless, degenerative changes that proceed at a rapid rate may result in localized segmental instability of the spine. Thus, abnormal translation and abnormal rotation may occur at the site of the segmental instability [15]. The lumbar spine is a particularly common place for the occurrence of this condition. When

conservative measures fail, surgical stabilization may be indicated, but only after a thorough assessment [15].

Disk Herniation

Although disk herniation can occur in any age group, the peak incidence of lumbar disk herniation appears to be in the

Figure 11-7

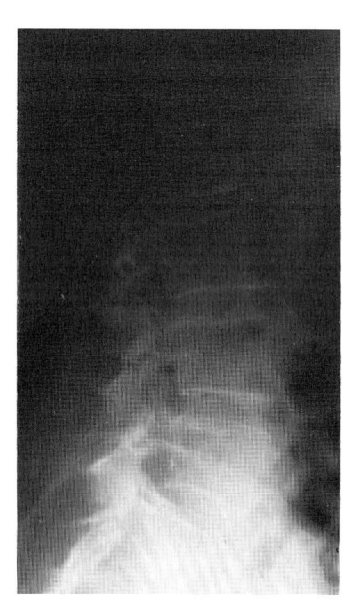

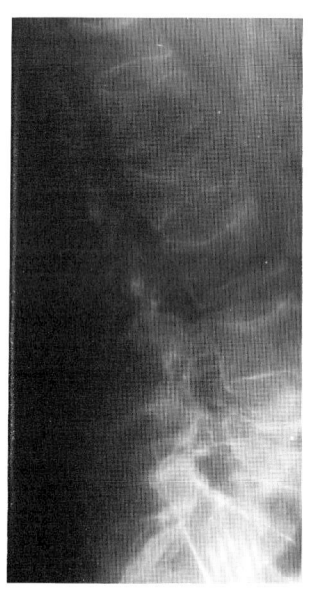

Two lateral x-rays of the lumbar spine on the same patient taken approximately 6 weeks apart. Note the marked decrease in bone density in the x-ray on the left. Observe the marked progression and the numerous vertebral compression fractures, which occurred over a relatively short time.

age range of 35 to 40 years [16]. The pathophysiology of disk herniation is more likely related to repetitive loading phenomena than to a single event. Symptoms of a disk herniation are often triggered by a relatively trivial twisting event. This trivial event results in disruption of the outer fibers of the annulus fibrosus. The resulting egress of disk material creates irritation and tension of the adjacent nerve root.

Spinal Stenosis

Lumbar spinal stenosis is a common clinical entity in which the vertebral canal is narrowed [17] (Fig. 11-8). This is usually a direct consequence of aging or degenerative changes in the spine [17, 18]. Even though most elderly adults probably would demonstrate varying degrees of vertebral canal stenosis on computerized tomography (CT) scans, most of these individuals do not have symptoms related to a stenotic spinal canal. In those individuals who do notice a decreased ability to ambulate and an adverse change in their quality of life, surgical decompression (removal of bone) of the stenotic segment of spine is often effective and satisfying for the patient [18]. The factors that influence the need for surgical intervention include the patient's subjective symptoms, physical examination, electromyographic findings, and the findings of lumbar myelography and CT scan. Spinal stenosis may occur as a central phenomenon whereby the structures of the cauda equina are all affected or as a relatively localized lateral root entrapment phenomenon in which the facet joint hypertrophies, entrapping the root at its egress from the vertebral canal [17, 18].

Figure 11-8

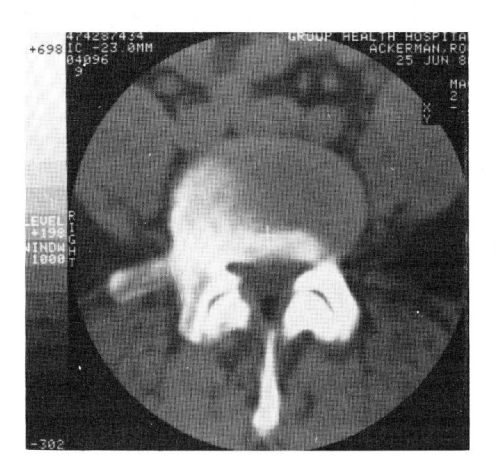

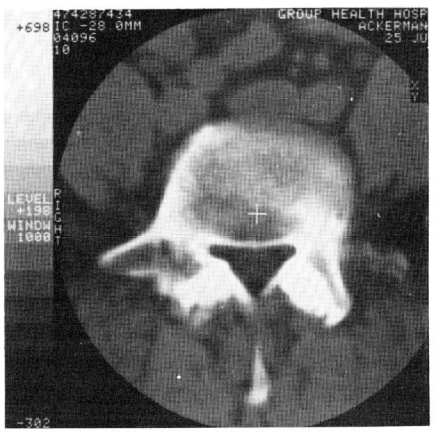

CT scan of an individual with lumbar spinal stenosis. Note the prominence of the facet joint in the recess. There is also an element of mild central stenosis.

PREVENTION AND TREATMENT

Disability, functional impairment, and pain are the symptoms that affect a large percentage of the adult population (approximately 80%) as a result of age-related changes in the spine. Both fundamental research into the components of the spine and clinically oriented research focusing on improved selection criteria and techniques for medical and surgical management of patients who suffer from age-related complaints are needed.

Preventing or minimizing age-related changes of the spine is a worthwhile goal. At present, however, little scientific evidence exists to guide the concerned health care professional. Changes in nutrition, stress, posture, and activity are areas now under investigation that eventually may lead to prevention or minimizing of spine-related symptoms.

REFERENCES

1. Valkenburg, H.A., Haanen, H.C. The Epidemiology of Low Back Pain. NIAMMD Workshop on Idiopathic Low Back Pain, Miami, Florida, December 1980.
2. Eyre, D.R. Biochemistry of the intervertebral disc. *Int. Rev. Connect. Tissue Res.* 8:227–291, 1979.
3. Pritzker, K.P. Aging and degeneration in the lumbar intervertebral disc. *Orthop. Clin. North Am.* 8:65–77, 1977.
4. Taylor, T.K.F., Akeson, W.H. Intervertebral disc prolapse: a review of morphologic and biochemic knowledge concerning the nature of prolapse. *Clin. Orthop.* 76:54, 1971.
5. Nachemson, A.L., Schultz, A.B., Berkson, M.H. Mechanical properties of human lumbar spine motion segments: influences of age, sex, disc level and degeneration. *Spine* 4:1–8, 1979.
6. Galante, J.O. Tensile properties of the human lumbar annulus fibrosus. *Acta Orthop. Scand. [Suppl.]* 100:1, 1967.
7. Maroudas, A., Nachemson, A., Stockwell, R., Urban, J. Some factors involved in the nutrition of the intervertebral disc. *J. Anat.* 120:113–130, 1975.
8. Mooney, V., Robertson, J. The facet syndrome. *Clin. Orthop.* 115:149–156, 1976.
9. Sullivan, J.D., Farfan, H.F. Pathological changes with intervertebral joint rotational instability in the rabbit. *Can. J. Surg.* 14:71, 1971.
10. Nachemson, A. Lumbar interdiscal pressure. *Acta Orthop. Scand. [Suppl.]* 43:1960.
11. Farfan, H.F. *Mechanical Disorders of the Low Back.* Philadelphia: Lea & Febiger, 1973.
12. Carter, D.R., Hayes, W.C. Bone compressive strength: the influence of density and strain rate. *Science* 194:1174–1176, 1976.
13. Burstein, A.H., Reilly, D.T., Martens, M. Aging of bone tissue: mechanical properties. *J. Bone Joint Surg. [Am]* 58A:82, 1976.
14. Carter, D.R., Schwab, G.H., Spengler, D.M. Tensile fracture of cancellous bone. *Acta Orthop. Scand.* 51:733–741, 1980.
15. MacNab, I. *Backache.* Baltimore: Williams & Wilkins, 1977.
16. Spangfort, E.V. The lumbar disc herniation. *Acta Orthop. Scand. [Suppl.]* 142:1–95, 1972.
17. Kirkaldy-Willis, W.H., Paine, K.W.E., Cauchoix, J., McIvor, G.W.D. Lumbar spinal stenosis. *Clin. Orthop.* 99:30–50, 1974.
18. Tile, M., McNeil, S.R., Zarins, R.K., et al. Spinal stenosis: results of treatment. *Clin. Orthop.* 115:104–108, 1976.

Current Concepts of the Pathogenesis of Osteoarthritis

Leon Sokoloff, M.D.

The classic pathologic studies of osteoarthritis by Heine [1] in 1925 and Bennett, Waine, and Bauer [2] in 1942 were carried out on specimens obtained at necropsy. They led to the concept that the disorder has its inception in the degeneration of articular cartilage. During the past two decades, a large amount of specimen material has become available to the surgical pathologist as a result of developments in reconstructive joint surgery. Together with our new knowledge of the biology of articular cartilage, this has led to a more complex picture of the nature of osteoarthritis.

What comes out of these studies is a series of striking analogies between the pathogenesis of osteoarthritis and atherosclerosis. Both are chronic, multifactorial disorders of the older population that have their onset years before they produce symptoms. The implications for management of each must accordingly be entirely different in the presymptomatic and symptomatic phases. Pharmacologic interventions in a patient who has suffered a myocardial infarction are hardly comparable to those one might entertain in a 10-year-old child who has anatomic lipid streaks in the coronary arteries. So it may be with degenerative joint disease. In both osteoarthritis and atherosclerosis, mechanical and biologic factors interact in the evolution of the lesions. Each is localized rather than diffuse. In each, there is considerable variation in localization, in the complexity of the character and composition of the lesions, and in their consequences. Common to both osteoarthritis and atherosclerosis, biological and biochemical oversimplifications have been invoked by single-minded investigators as the key to the pathogenetic process.

OSTEOARTHRITIS AS A REMODELING PROCESS

Osteoarthritis is an inherently noninflammatory deformity of movable joints characterized by two discrete pathologic processes: (a) deterioration and detachment of the bearing surface, and (b) proliferation of new osteoarticular tissue at the margins and beneath the detached joint surface. The sequences of these events are difficult to sort out because the process is ongoing and self-procreating. Three general concepts of these sequences have their own constituencies but seem to be overly simplistic:

1. *Osteoarthritis is a degeneration of articular cartilage that leads progressively to denudation of the joint surface.* If this were valid, little or no remodeling of the bone should occur. Only rarely is this the case, however, and usually it is the consequence of antecedent inflammation or metabolic peculiarity in which mechanical abnormalities are absent.
2. *Osteoarthritis begins as fibrillation of cartilage that leads to secondary remodeling of the bony components of the joint.* This is the common view. A principal difficulty is that one is hard put to isolate any individual finding as a unique morphologic event that precedes others in the complicated changes seen in histologic sections. This view does not, for example, take into account the remodeling of the osteochondral junction as an early age-related change in cartilage.
3. *Osteoarthritis is the consequence of changes in the stiffness of subchondral bone.* This view suffers the same limitations as the preceding one.

It seems unrealistic to attempt to identify a unique initial event in the osteoarthritic process. The structural disintegration of the osteoarticular junction and abrasion lead to the loss of substance of the articular surface. They also are responsible for the proliferative phenomena, including the formation of new cartilage at the surface of the osteoarthritic joint.

Remodeling is the alteration of the internal and external architecture of the skeleton as dictated by Wolff's law in response to variation in mechanical loading. It involves removal of bony tissue at certain points while it is being laid down elsewhere. The concept applies also to changes in the shape of the joint with age and osteoarthritis. Remodeling of articular cartilage is one major aspect of the remodeling joint in osteoarthritis and has three principal components: (a) structural removal of matrix, (b) formation of new cartilage, both its cells and matrix, and (c) endochondral calcification and ossification. These events are intimately associated with disturbances of mechanical forces acting on the joint surface. Noteworthy advances have taken place in our knowledge of the individual cellular and biochemical processes in the remodeling of cartilage. Neither the biomechanical nor the cellular events can be ignored in evaluating pharmacologic interventions in the management of degenerative joint disease.

ETIOLOGY

The two principal etiologic factors commonly invoked are aging and mechanical insult. It is to the latter particularly that the distinction between primary and secondary forms is most relevant. Secondary osteoarthritis refers to those instances that supervene on preexisting damage to joints, inflammatory or noninflammatory. Primary osteoarthritis is the form in which no traumatic or other predisposing articular etiology can be assigned. Here intrinsic degeneration of the articular tissue is presumed to underlie the development of the disease. In practice it is difficult in most instances to distinguish on anatomic grounds between the two types of osteoarthritis.

There are great divergences of opinion about the relative frequencies of these types. At one pole is the view that subtle roentgenographic features identify the vast majority of even idiopathic osteoarthritis of the hip as being of secondary type [3]. The localization of the areas of greatest joint space narrowing (e.g., superolateral or medial) and the configuration of osteophytes have been proposed as guides to a particular antecedent etiologic abnormality, both on clinical x-ray films [4] and in excised specimens [5]. In surgically resected femoral heads, the changes are usually so far advanced and diverse that these interpretations are not easily sustained. These presently unresolved matters obviously must enter into concepts that place primary emphasis on metabolic or inflammatory rather than biomechanical bases to the development of degenerative joint disease.

Senescence as an etiologic factor can mean two quite different things. It may simply represent the accretion of a series of cumulative insults to the articular tissue or, more biologically, time-dependent molecular alterations that take place in the cartilage independent of acquired lesions. In articular cartilage, a long-protracted, low-grade thermal degradation of the collagen or interaction of the collagen with metabolites, such as certain aldehydes, might represent such a biologic aging of cartilage. Dehydration of cartilage, as has been proposed in the case of spondylosis deformans, does not occur as a progressive phenomenon of aging in the articular cartilage. Chemical alterations presumably would change the material properties of the cartilage upon which its functional integrity depends.

One way in which the contribution of aging might be assessed is to compare the severity of the clinically obtrusive lesions with the changes found in a general aging population. The limited available data [6, 7] indicate that the deterioration of the joint surfaces progresses linearly with age. The slope is greater in the patellofemoral than in the hip joint. The changes in surgically resected specimens fall far outside the scatter in the natural history of the aging hip. This suggests that some local factor aside from aging is of major etiologic importance, at least in the hip [8].

REPAIR OF ARTICULAR CARTILAGE

The cornerstone of the wear-and-tear theory of osteoarthritis is the concept that articular chondrocytes cannot replicate themselves. It is presently clear, however, that two modes of repair are in fact biologically possible.

Aside from histologic evidence of clonal proliferation (so-called brood capsule formation) in fibrillated cartilage, in vitro culture documents unequivocally the ability of adult articular chondrocytes to divide. In explant culture, synthesis of DNA occurs both in situ and in outgrowth under controlled conditions. Partial lysis of the matrix is a sine qua non for cell proliferation [9].

Formation of cartilage also comes about through metaplasia and remodeling of noncartilaginous subchondral and marginal connective tissue of the damaged joint surface. The histogenetic reactions are entirely comparable to callus formation that occurs in fractures. Osteophytes of this nature and, indeed, the deformity of osteoarthritis are in effect largely an aberrant repair reaction.

VARIANT FORMS OF OSTEOARTHRITIS

Minor degenerative changes are widespread in joints at necropsy. Whether there are forms of symptom-producing osteoarthritis in multiple joints remains controversial. The association of Heberden's nodes with osteoarthritis of the hip has been affirmed by some [10] and denied by others [11]. In humans, and indeed in subhuman species, there often is a dichotomy between isolated osteoarthritis of the hip and osteoarthritis that affects many joints but spares the hips [12]. The relationship of generalized osteoarthritis to diffuse idiopathic skeletal hyperostosis (DISH) is now coming under scrutiny [13, 14].

Aside from overtly erosive forms of Heberden's nodes, histologic foci of chronic nonspecific synovitis are commonly seen in joints resected for osteoarthritis [15]. The lesions are usually mild and sometimes accompanied by foreign body giant cells around joint detritus. Considerable villous thickening by fibrous tissue is often seen, as well as small deposits of hemosiderin. An extreme hypothesis, proposed by Ehrlich [16], suggests that all osteoarthritis may result from antecedent inflammation. Intracellular and some extracellular components of immune complexes have been reported in synovium of osteoarthritic hips [17]. Cooke [18] has found immune complex components in the tangential layer of articular cartilage frequently in primary but not in secondary coxarthrosis. He has therefore suggested that these are two distinct entities and that immunologic abnormalities may be part of generalized osteoarthritis. A priori, the synovitis in osteoarthritis may be a secondary phenomenon. That it may arise as an autoimmune response to degraded joint detritus is suggested by the occasional presence of lymphokines in the synovial fluid in osteoarthritis [19]. There also is some experimental support for this view [20].

Minute quantities of hydroxyapatite [21] and calcium pyrophosphate dihydrate crystals [22, 23] have repeatedly been found in synovial fluid in osteoarthritis. These observations have aroused much interest in abnormal mineral metabolism in the etiology of certain types of osteoarthritis. The possibility that the crystals may evoke synovitis has been suggested [24]. Pathologists usually do not observe chondrocalcinotic deposits in the cartilage.

Synthesis of hydroxyapatite and calcium pyrophosphate is an integral part of the mineralization that accompanies the remodeling of bone and cartilage in osteoarthritis [22, 25]. The synovial fluid crystals may thus represent a manifestation rather than a cause of the degenerative process. This does not exclude the possibility that the crystals contribute to a perpetuation of the process.

BASIS OF CLINICAL COMPLAINTS

By the time patients present with serious complaints from osteoarthritis, the lesions commonly are quite far advanced. Osteophytosis of the knee [26], hip [27], or vertebrae per se is a poor guide to the complaints or prognosis. Development of osteophytes in the course of osteoarthritis is another matter. The location, size, and concomitants of osteophytes vary widely. Patellar spurs extending into the quadriceps tendon in diffuse idiopathic skeletal hyperostosis have a quite different significance and character from the osteophytes in osteoarthritis of the knee. Teleologically, marginal osteophytes increase the area of the surface available for supporting the load on joints. A diminution of pain during the course of their evolution has been attributed to this. A price is paid, however, in that the increasing irregularity of the joint impairs its mobility. In other circumstances where the osteophyte impinges on nerves, the consequences with respect to pain are different. Pain also has other sources. A major cause must be the microfractures of the articular cortex that lead to the deformity, pseudocyst formation, and venous congestion so common in the osteoarthritic bone. Another component is the capsular reaction, both mechanical tears and the inflammatory processes already noted. Musculoligamentous imbalances and psychosocial factors must also fit into the equation.

CONCLUSIONS

Osteoarthritis is not so much a single disease as a pattern of reaction of articular tissues to mechanical and biologic events. It is in this context that one attempts to discriminate certain subsets. If one strains the analogy to atherosclerosis, there is one end of the spectrum in which metabolic peculiarities dominate: ochronosis and certain chondrodystrophies in osteoarthritis, Hurler's disease or hyperbetalipoproteinemia in coronary heart disease. At the other extreme is the biomechanically dominated lesion: the dysplastic hip in osteoarthritis, severe hypertension in atherosclerosis.

There is a recrudescence of interest in generalized patterns of osteoarthritis as distinct from those that affect single joints. In the former, complaints are more often referable to the knees and hands than confined to the hips. When the hip is affected, it has a more concentric than a migratory pattern. Inflammatory manifestations are more common in the generalized varieties. It is in this group that the role of immune complexes and crystal deposition is currently the center of attention.

Osteoarthritis is not the expression of an inability of cartilage to repair but an aberration of the repair process. Many attempts at pharmacologic management of osteoarthritis have been directed at the metabolic preservation or restoration of articular cartilage. They have not taken into account the reparative nature of the osteoarthritic lesion and may conceivably carry a potential for doing more harm than good. This is true of compounds such as Rumalon [28], fibroblast growth factor [29], Arteparon [30], and other antiproteases [31].

Repair of cartilage is biologically permissible. The character and degree of repair are governed by the mechanical environment in which it occurs.

An area of research that deserves more systematic attention is the transduction of mechanical into biologic signals for the growth and differentiation of articular mesenchyme. How does cartilage know when to become calcified and when not to be? What governs the differentiation of reparative callus into hyaline cartilage rather than fibrocartilage—or for that matter, bone or bone marrow—all of which are components of the osteophyte?

At a clinical level, this sort of information would let us know the bounds on the ability of the articular surface to remodel and the osteoarthritic lesions to be reversed. One is interested in the interactions of physical forces and the formation of cartilage, as Salter and colleagues [32] have studied in the repair of experimental defects in joint surfaces. The ultimate logic of this sort of thinking will be the proper mix of physical and biologic interventions in future generations of reconstructive joint surgery.

REFERENCES

1. Heine, J. Über die Arthritis deformans. *Virchows Arch. [Pathol. Anat.]* 260:521–663, 1926.
2. Bennett, G.A., Waine, H., Bauer, W. *Changes in the Knee Joint at Various Ages.* New York: Commonwealth Fund, 1942.
3. Stulberg, S.D., Cordell, L.D., Harris, W.H., et al. Unrecognized childhood hip disease: a major cause of osteoarthritis of the hip. In *The Hip. Proceedings of the 3rd Open Meeting of the Hip Society.* St. Louis: C.V. Mosby, 1974, pp. 212–228.
4. Gofton, J.P. Studies in osteoarthritis of the hip. *Can. Med. Assoc. J.* 104:679–683, 791–799, 911–915, 1971.
5. Resnick, D. Patterns of migration of the femoral head in osteoarthritis of the hip: roentgenographic-pathologic correlation and comparison with rheumatoid arthritis. *AJR* 124:62–74, 1975.
6. Meachim, G., Emery, I.H. Quantitative aspects of patello-femoral cartilage fibrillation in Liverpool necropsies. *Ann. Rheum. Dis.* 33:39–47, 1974.
7. Byers, P.D., Contepomi, C.A., Farkas, T.A. A post mortem study of the hip joint: including the prevalence of the features of the right side. *Ann. Rheum. Dis.* 29:15–31, 1970.
8. Sokoloff, L. Aging and degenerative diseases affecting cartilage. In Hall, B.K. (ed.), *Cartilage,* vol. 3. New York: Academic Press, 1983, pp. 109–141.
9. Krystal, G., Morris, G.M., Sokoloff, L. Stimulation of DNA synthesis by ascorbate in cultures of articular chondrocytes. *Arthritis Rheum.* 25:318–325, 1982.
10. Dequeker, J., Brussens, A., Creytens, G., Bouillon, R. Ageing of bone: its relation to osteoporosis and osteoarthritis in postmenopausal women. *Front. Horm. Res.* 3:116–130, 1975.
11. Yazici, H., Saville, P.D., Salvati, E.A., et al. Primary osteoarthrosis of the knee or hip: prevalence of Heberden nodes in relation to age and sex. *J.A.M.A.* 231:1256–1260, 1975.

12. Huskisson, E.C., Dieppe, P.A., Tucker, A.K., Cannell, L.B. Another look at osteoarthritis. *Ann. Rheum. Dis.* 38:423–428, 1979.
13. Utsinger, P.D., Resnick, D., Shapiro, R. Diffuse skeletal abnormalities in Forestier disease. *Arch. Intern. Med.* 136:763–768, 1976.
14. Lagier, R., Baud, C.A. Diffuse enthesopathic hyperostosis: anatomical and radiological study on a macerated skeleton. *Fortschr. Röentgenstr.* 129:588–597, 1978.
15. Ito, S., Bullough, P. Synovial and osseous inflammation in degenerative joint disease and rheumatoid arthritis of the hip: a histometric study. *Trans. Orthop. Res. Soc.* 4:199, 1979.
16. Ehrlich, G.E. Osteoarthritis. *Arch. Intern. Med.* 138:688–689, 1978.
17. Pringle, J.A., Byers, P.D., Brown, M.E.A. Immunofluorescence in osteoarthritis. *Nature* 274:294, 1978.
18. Cooke, T.D.V. The interactions and local disease manifestations of immune complexes in articular collagenous tissues. *Stud. Joint Dis.* 1:158–200, 1980.
19. Stastny, P., Rosenthal, M., Andreis, M., Ziff, M. Lymphokines in the rheumatoid joint. *Arthritis Rheum.* 18:237–243, 1975.
20. Champion, B.R., Poole, A.R. Immunity to homologous cartilage proteoglycans in rabbits with chronic inflammatory arthritis. *Collagen Res.* 1:453–473, 1981.
21. Dieppe, P.A., Huskisson, E.C., Willoughby, D.A. The inflammatory component of osteoarthritis. In Nuki, G. (ed.), *The Aetiopathogenesis of Osteoarthrosis.* Kent, England: Pitman Medical, 1980, pp. 117–122.
22. Howell, D.S., Muniz, O., Pita, J.C., Enis, J.E. Extrusion of pyrophosphate into extracellular media by osteoarthritic cartilage incubates. *J. Clin. Invest.* 56:1473–1480, 1975.
23. Silcox, D.D., McCarty, D.J., Jr. Elevated inorganic pyrophosphate concentrations in synovial fluids in osteoarthritis and pseudogout. *J. Lab. Clin. Med.* 83:518–531, 1974.
24. Schumacher, H.R., Miller, J.L., Ludivico, C., Jessar, R.A. Erosive arthritis associated with apatite crystal deposition. *Arthritis Rheum.* 24:31–37, 1981.
25. Ali, S.Y. Mineral-containing matrix vesicles in human osteoarthrotic cartilage. In Nuki, G. (ed.), *The Aetiopathogenesis of Osteoarthrosis.* Kent, England: Pitman Medical, 1980, pp. 105–115.
26. Ahlbach, S. Osteoarthrosis of the knee: a radiographic investigation. *Acta Radiol. [Suppl.] (Stockh.)* 227:7–72, 1968.
27. Danielsson, L.G. Incidence and prognosis of coxarthrosis. *Acta Orthop. Scand. [Suppl.]* 66:1–114, 1964.
28. Denko, C.W. Restorative chemotherapy in degenerative hip disease. *Agents Actions* 8:268–279, 1978.
29. Wellmitz, G., Petzold, E., Jentzach, K.D., et al. The effect of brain fibroblast growth factor activity on regeneration and differentiation of articular cartilage. *Exp. Pathol. (Jena)* 18:282–287, 1980.
30. Bach, G.L., Panse, P., Zeiller, P. Glykosaminoglykanpolysulfat (GAGPS, Arteparon) zur Basistherapie der Arthrose: III. Biochemischdiagnostische und klinische Untersuchungen zur intramuskulären Anwendung von GAGPS. *Z. Rheumatol.* 36:269–274, 1977.
31. Telhag, H. Effect of tranexamic acid (Cyklokapron) on the synthesis of chondroitin sulphate and the content of hexosamine in the same fraction on normal and degenerated joint cartilage in the rabbit. *Acta Orthop. Scand.* 44:249–255, 1973.
32. Salter, R.B., Simmonds, D.F., Malcolm, B.W., et al. The biological effect of continuous passive motion on the healing of full-thickness defects in articular cartilage: an experimental investigation in the rabbit. *J. Bone Joint Surg. [Am]* 62A:1232–1251, 1980.

Early Aging Nutritional Changes at the Base of the Articular Cartilage

Darrel W. Haynes, M.S., Ph.D.

It has been demonstrated, both clinically and experimentally, that sepsis or trauma, or both, can cause osteoarthritis. However, no single causative agent has been identified in the etiology of the idiopathic osteoarthritis that has often been associated with age.

Osteoarthritis has often been thought of as part of the natural aging process. It has been thought that the cartilage wears at the surface until it is so thin that it fibrillates and is worn down to the underlying bone. However, if the articular cartilage on both surfaces of the joint were to wear, the joint would become incongruent and only a point contact between the surfaces would exist. This is not so, as it has been shown that the joint becomes more congruent with age. No explanation has been given to date for this phenomenon of increased congruence, as articular cartilage is incapable of significant cellular division and repair in normal circumstances.

Most of the studies of human osteoarthritis have been based on observations of pathologic joints taken either at autopsy or at surgery. In attempting to identify the etiology of the disease from these deformed joints, one can make the analogy of trying to study the anatomy of an insect after it has been squashed. The joints are often so deformed that the original 3 to 5 mm of thickness of cartilage, or the original joint line, has been destroyed. Many of the studies using autopsy material have shown that as the population increases in age one will find more osteoarthritic joints. However, it cannot be explained why one finds a perfectly normal, healthy joint in a 100-year-old person and a diseased joint in a 30-year-old person.

Therefore, since the joint has failed because of cartilage degeneration, a single causative factor may be the lack of metabolic activity caused by an interruption of its nutritional route. It has long been known that articular cartilage does not contain blood vessels, and the mechanisms by which nutrients reach the cartilage have been a problem of continued interest. There are three theoretical mechanisms by which nutrients may enter hyaline articular cartilage, namely, from the synovial joint fluid alone, from the bone marrow alone, or from both directions. Recent discussion in the literature has resolved itself into an argument as to whether the route from the bone marrow exists in practice or not. Greenwald and Haynes [1], utilizing a fluorescent-dye method on cadaveric femoral heads, demonstrated diffusion of dye from the marrow into articular cartilage through defects in the subchondral plate. These defects are occupied by vascular channels and marrow tissue, as described by Woods, Greenwald, and Haynes [2]. Additional evidence for the existence of such defects has recently been obtained by scanning electron microscopy by Mital and Millington [3].

The view that the adult subchondral plate is not vascularized is based on observation of experimenal animal skeletons, mainly of rabbits by Hodge and McKibbin [4] and Honner and Thompson [5].

Criticism of the use of human cadaveric material because of the possible effect of autolytic changes has come from McKibbin and Maroudas [6], who have also expressed the opinion that, although subchondral vascular channels exist, they are too few in number to be of significance in the nutrition of articular cartilage.

Although it is generally agreed [1, 4, 5] that immature joints receive their nutrition from both the synovial joint fluid and bone marrow, the major disagreement is the exact mechanism in the adult human. My preliminary observations of the adult articular surfaces other than the femoral head have shown some striking differences in the number and location of subchondral vessels in different articulations.

The following experiment should demonstrate the role of the subchondral vessels in the nutrition of adult articular cartilage.

EXPERIMENTATION

To avoid further controversy concerning the extrapolation of data from animals to humans, it was decided to study only the freshly amputated lower limbs from young adult patients. Three hindquarter and two above-knee amputations were received in the laboratory within a few minutes after their removal. The external iliac or femoral artery, as appropriate, was cannulated and the limb perfused with 20% pyranine in normal saline at a pressure of 120 mm Hg. The fluorescent dye pyranine was selected as an indicator because in fresh sections, stained by immersion, the dye bound itself evenly throughout the cartilage. Extremely small amounts of the dye were easily detectable because of its intense fluorescence. The bone and calcified cartilage displayed no affinity for the dye, whereas osteoid and cartilage did. It was concluded that the dye bound itself to collagen provided the latter was not calcified. The dye was readily soluble in aqueous solutions, and therefore, after initial penetration, it was necessary to maintain the tissue in a frozen state to avoid seepage of dye into unstained areas. Freezing the tissues did not produce architectural artifacts detectable by light microscopy. The dye has a molecular weight of 364, which is greater than that of known amino acids in collagen. It is assumed that if the large dye molecule can pass through, so too could the smaller nutritional amino acids. The perfusion time ranged from 1 to 5 minutes.

When perfusion was completed, slices were taken from the femoral condyle and the patella along with complete metatarsal heads and phalanges and immersed in liquid nitrogen. The greatest time interval between the amputation and freezing of the specimen was 30 minutes.

While still frozen, the specimens were sliced on a band saw to approximately 5 mm thick and were then attached to metal chucks with ice. The chucks were inserted into a cryostat containing a Cambridge rocking microtome, and 15-μ sections were cut. The sections were attached to fluoro-free slides and coverslips secured with D.P.X. mountant. The sections were examined on a Zeiss photomicroscope using ultraviolet light with a BG 3/4 exciter filter and a 50/44 barrier filter. Tungsten polarized light was also used to identify the precise location of the dye.

RESULTS

Within seconds of the beginning of the perfusion the skin of the whole specimen became green from the fluorescent dye. Histologic sections from joints of the limb perfused for 5 minutes revealed fluorescence of the full thickness of the articular cartilage. For this reason, the other limbs were perfused for only 1 or 2 minutes so that the dye could be observed as it was just entering the cartilage. In these short perfusion specimens, the following conclusive observations were made.

In the femoral condyle and patella, fluorescence was present in the superficial surface of the cartilage to a depth of about 500 µ. However, there was also fluorescence in the deep zone that was separate from the superficial surface staining. The greatest intensity of the basal fluorescence was seen around the vascular channels, which were most numerous beneath the weight-bearing areas. One could actually visualize the dye leaving the channels and perfusing into the articular cartilage. The volume of dye entering the cartilage was greater at the base than at the surface.

In the first metatarsal phalangeal joint, the superficial part of the cartilage of the metatarsal head was stained uniformly. However, fluorescence of the basal zone was only observed inferiorly, where cartilage makes contact with synovium supported by densely collagenized capsular tissue. The basal fluorescence was most intense around the vascular channels, which were apparently confined only to the plantar third of the subchondral plate. Both the superficial and deep zones of the cartilage of the phalanx were stained.

In the fifth metatarsal phalangeal and all interphalangeal joints, only the superficial zone of the cartilage showed fluorescence.

These results finally and conclusively show that in all adult human joints, nutrients can be received from the synovial joint fluid. However, where the joints contained subchondral vascular channels, the nutrients could enter the basal cartilage and diffuse into the cartilage as easily as they could diffuse into the superficial cartilage from the synovial fluid. Where there were no subchondral vessels, the cartilage relied entirely on the synovial fluid for its source of nutrition.

DISCUSSION

Herein appears to be the reason for the opposed views of those who have studied human and animal joints. The femoral head of the skeletally mature rabbit, which was the animal most investigators had studied, does not have subchondral blood vessels and therefore does not have a subchondral nutritional route. My limited observations have also revealed that there are no subchondral vessels in the femoral head of the dog, cow, goat, or sheep. The only mature animal that appears to parallel the human is, in fact, the primate.

In observing microscopically the subchondral areas of other adult human joints, one can see the presence or absence of vascular channels. It would appear that in all the larger joints there are subchondral vessels, whereas in the phalanges there are none. The number of vessels was always greatest in the center of the joint. The center of the joint is normally the thickest layer of cartilage and is also the area of greatest weight bearing. It may be postulated that the greater thickness and weight-bearing cartilage will require an additional nutritional route, as they are also the greatest distance away from the synovium and the source of joint fluid. Any breakdown in this subchondral nutritional mechanism could lead to the degradation of the cartilage because of poor metabolic activity.

It has been shown by Woods, Greenwald, and Haynes [2] that these subchondral vessels decrease with age. In further experiments, I have been able to show that the subchondral plate advances toward the articular surface, calcifying the cartilage at its base [7]. As this advance occurs, the subchondral vascular channels are sealed off, therefore reducing the nutrients that are able to penetrate into the base of the articular cartilage. Calcification is a natural response to dead or dying tissue, and it is not known if the calcification occurs because the nutritional route has been impaired or whether some other metabolic phenomenon excites the calcification, thus walling off the nutrient vessels and causing the demise of the cartilage.

The early confusion about joint nutrition was caused by the first investigators' confusing the difference in nutrition between mature and immature animals. Once this was recognized, they then declared that adult cartilage does not have a subchondral route. However, they did not look at the adult human cartilage and observe that there was a mechanism by which nutrients could reach the base of the articular cartilage through vascular channels. This free extrapolation from the animal condition to the human, in this case, has been a great error. It is clear that most adult human joints, other than the phalangeal joints, have and therefore must require a subchondral route. It is interesting to note that these subchondral vessels decrease with age and that, in contrast, the incidence of osteoarthritis increases with age. It also may be that there is a type of stroke in the subchondral area, thus causing an infarct of the articular cartilage, which may then calcify, thus reducing the cartilage thickness without altering its surface. This thin cartilage may then be subjected to mechanical wear and destruction.

CONCLUSIONS

Adult articular cartilage is supplied by a synovial and subchondral route in most joints, with the exception of the phalanges. The importance of the subchondral route and the effect its depletion may have on the integrity of articular cartilage are unknown. Animal experiments, which have led to confusion and misunderstanding in the area of nutrition, should always be carefully evaluated before extrapolating results to humans.

REFERENCES

1. Greenwald, A.J., Haynes, D.W. A pathway for nutrients from the medullary cavity to the articular cartilage of the femoral head. *J. Bone Joint Surg.* [*Br*] 51B:747, 1969.
2. Woods, C.G., Greenwald, A.S., Haynes, D.W. Subchondral vascularity in the human femoral head. *Ann. Rheum. Dis.* 29:138, 1970.
3. Mital, M.A., Millington, P.F. Osseous pathway of nutrition to articular cartilage of the human femoral head. *Lancet* 1:842, 1970.
4. Hodge, J., McKibbin, B. The nutrition of mature and immature joint cartilage in rabbits. *J. Bone Joint Surg.* [*Br*] 51B:140, 1969.
5. Honner, R., Thompson, R.C. The nutritional pathways of articular cartilage. *J. Bone Joint Surg.* [*Am*] 53A:742, 1971.
6. McKibbin, B., Maroudas, A. Nutrition and metabolism. In Freeman, M.A.R. (ed.), *Adult Articular Cartilage*. London: Pitman, 1979, pp. 461–486.
7. Haynes, D.W. The mineralization front of articular cartilage. *Metab. Bone Dis. Related Res.* 25:55–59, 1980.

Biomechanics of Joint Deterioration and Osteoarthritis

▶

Eric L. Radin, M.D.
R. Bruce Martin, Ph.D.

Degenerative joint disease is one of the major gerontologic health problems that challenge medicine as an increasing proportion of our population becomes elderly. About 40 million adults in the United States have radiographic evidence of osteoarthritis in their hands and feet. Eighty-five percent of those over 70 years old show such effects in their spinal columns, so that the wearing out of joints is an almost universal phenomenon. Fortunately, most of this wear is asymptomatic, but in some cases pain and loss of function are severe. An estimated 175,000 Americans over age 65 are severely crippled with osteoarthritis [1].

It should be pointed out that while osteoarthritis is certainly an age-related disease, it is by no means exclusively a disease of aging. Osteoarthritis causes more absenteeism than any other form of joint pathology [1]. Thus, while osteoarthritis would appear to be a function of wear, it can occur early in life if the body's protective mechanisms are not functional or if such mechanisms are not able to cope with the severity of abuse that some of us heap upon our joints.

The term *osteoarthritis* apparently was first applied in 1889 by J.K. Spender. It has been used to describe many different forms of joint disease, including that due to mechanical wear. We prefer to call this joint wear *osteoarthrosis*. This preciseness of terminology is subtle but important. Arthritis means "joint inflammation"; in many forms of joint disease, inflammation is the primary pathologic problem, but in the form being addressed here, inflammation is secondary to mechanical failure. Therefore, we prefer to deemphasize the inflammatory aspect of the problem by using the term *osteoarthrosis* rather than *osteoarthritis*.

THE ROLE OF BONE IN JOINT FUNCTION

Whichever term you prefer, *osteoarthrosis* or *osteoarthritis,* it will have the prefix "osteo-," which is fortunate because the etiology of the disease has as much to do with bone as it does with cartilage. The purpose of an articular joint is to provide for simultaneous articulation and load transmission between one body segment and another. This function is intimately related to another function of the musculoskeletal system, however, and that is the dissipation of energy received through impacts with the environment. Thus, walking or running involves not only articulation under load, but also intraarticular loads that are several times body weight and large enough to cause trauma if they were not attenuated.

There are two major ways in which these impulsive loads are attenuated. The first is by muscle lengthening under tension [2], an active physiologic response. The second means of shock absorption is passive: deformation of the connective tissues carrying the load. These include cartilage, trabecular bone, cortical bone, tendons, and other tissues.

For some time it was thought that cartilage, being much more compliant and viscoelastic than bone, was the major shock absorber in a joint. Later, however, it was realized that the layer of cartilage is so thin that its total capacity for deformation and energy absorption is much less than that of the bone beneath it. This is clearly shown in Figure 14-1, where a load is applied to a typical articular joint. The load could, for example, be due to the foot striking the ground while running. The energy transmitted to the body in this activity will be partially stored in the skeletal tissues by virtue of their deformation, and the ability of each tissue to receive such energy depends inversely on its stiffness. The stiffness of articular cartilage is about 50 kg/mm^2, while those of cortical and trabecular bone are about 1000 and 9 kg/mm^2, respectively [3]. When one takes into account the areas over which the load is distributed in the cartilage, subchondral bone, and diaphysis, and the thickness of each of these regions, one finds that the subchondral bone absorbs 175 times more energy than does the cartilage. Thus, it is very clear that the subchondral bone plays a major role in absorbing mechanical energy and that this role hinges on its relatively low stiffness.

The other major role of the subchondral bone also is very much a function of its compliance. In order for the stress in the cartilage to be minimized when a force is transmitted across the joint, the joint surfaces should contact one another over as broad an area as possible. This is best achieved if the bone supporting the cartilage is compliant

Biomechanics of Joint Deterioration

enough to permit the cartilage of one side to mold itself to that of the other under various degrees of loading. Of course, this deformability must be elastic to allow the joint to return to its unloaded conformation.

Subchondral bone is trabecular in its structure, so its stiffness depends on the size and interconnections of the trabecular plates and struts. In general, these trabeculae tend

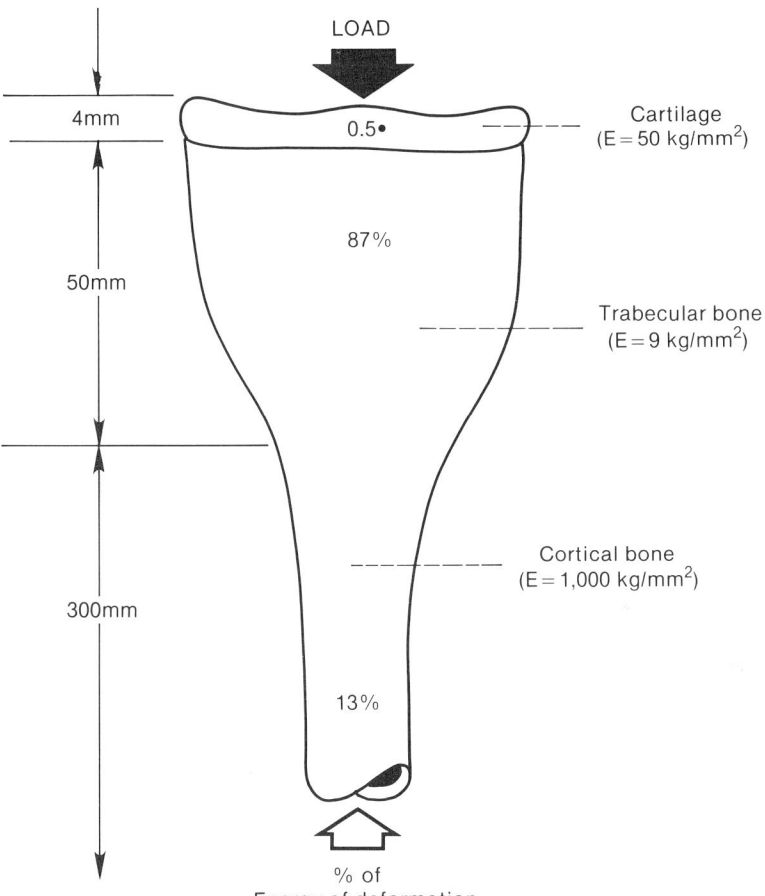

Figure 14-1

Schematic model for articular joint showing three regions of varying stiffness (or elastic modulus) (E) and geometry. The percentage of the total energy of deformation falling in each region is shown.

to be oriented along the directions of compressive and tensile stress. This is thought to minimize trabecular bending and shear stresses, as these are more likely to cause microfractures of the cancellous bone. The stiffness of cancellous bone has been shown to be inversely related to its porosity (the volume fraction occupied by soft tissues) and directly related to the amount of connectivity (i.e., bracing) between the trabeculae [4]. Furthermore, the development of osteoarthrosis can be shown to be related to local changes in these two variables [5]. This is readily understandable, since changes in the stiffness of the subchondral bone would affect both energy absorption and joint congruence, and both these effects could increase the stress within the cartilage.

STIFFNESS GRADIENTS AND CARTILAGE DAMAGE

All joints, because of their morphology, tend to have certain articular surfaces that bear a disproportionate share of the joint's load. The subchondral bone beneath these habitually loaded areas will experience greater stresses than the surrounding bone and will remodel according to Wolff's law. Its porosity will decrease, it will become highly interconnected, and, consequently, it will become more stiff than the bone beneath the habitually nonloaded areas of the joint. One universally finds fissuring or fibrillation of the cartilage over the nonloaded areas of joints and particularly at the margin between loaded and nonloaded areas [6]. It has been suggested that fibrillation over nonloaded areas is due to poor nutrition, the cartilage requiring cyclic loading to pump fluids through its structure. Thaxter and associates [7] have shown, however, that in a statically loaded joint, the unloaded peripheral cartilage enjoys persistent good health, suggesting that diffusion rates through articular cartilage are high enough to allow for adequate nutrition of the chondrocytes without mechanical pumping. An alternative explanation for cartilage fibrillation in these areas is that it is initiated over areas of stiffness gradients in the underlying bone [8]. Stress is "concentrated" over such discontinuities, and, in addition, these regions of articular cartilage would experience higher shear stresses. This is shown graphically in Figure 14-2, where the cartilage is seen to be "scissored" over the edge of the stiffer bone. Like most materials, cartilage is weaker in shear than in tension or compression [3]. Therefore, concentrated shear stresses at the margins between habitually loaded and nonloaded areas may well be responsible for cartilage fibrillation over these areas.

This fibrillation in habitually nonloaded areas is usually asymptomatic, seems to be unavoidable, and probably represents a relatively mild form of age-related wear [9]. It does not usually progress to cartilage loss, eburnated bone, and osteoarthrosis. Indeed, experimental dicing up of articular cartilage in animals does not lead to any osteoarthrosic changes [10]. What, then, *does* cause osteoarthrosis?

DEVELOPMENT OF OSTEOARTHROSIS

The primary etiologic factor behind osteoarthrosis is an imbalance between the mechanical stress on the joint and the ability of the joint tissues to withstand that stress. Stress concentrations are the principal cause of such an imbalance. These stress concentrations result from joint incongruity and bony remodeling. Joint incongruity can be primarily mor-

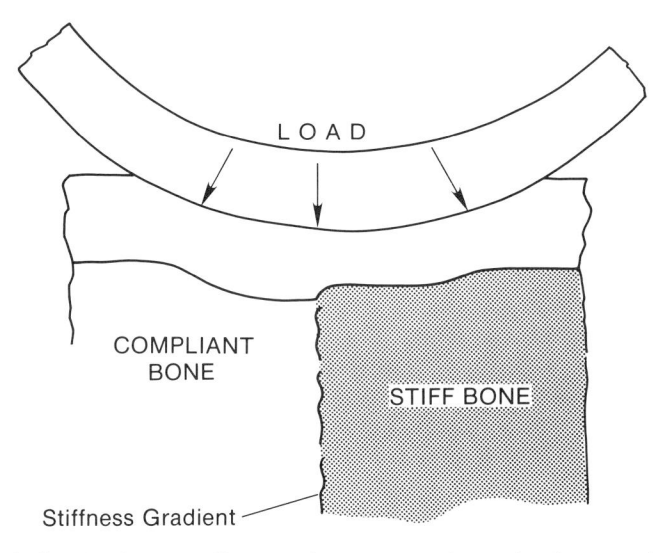

Figure 14-2

Sketch showing how a stiffness gradient in the subchondral bone produces a shearing effect in the overlying cartilage when it is loaded by the opposing articular surface.

phologic and due to congenital, developmental, or traumatic events. On the other hand, it can also be caused by progressive stiffening of the subchondral bone so that the joint surfaces no longer conform under load. This gradual change is the result of the healing of microfractures in the trabecular bone over many years. Bone, like all other structural materials, is subject to fatigue damage, and in spongy bone this takes the form of fractures of individual trabeculae. These fractures heal by a miniature version of diaphyseal fracture healing; a callus develops around the fractured trabeculae, calcifies, and is slowly remodeled away to leave a new trabeculum.

The accumulation of fatigue damage depends on the frequency and magnitude of the applied loads, and if either of these factors is too great, the repair process will not be able to keep up with the rate of fracture. In the worst cases this leads to a clinical fatigue fracture, but in milder cases there is simply not enough time for the trabecular calluses to be remodeled away. This results in a gradual increase in the stiffness of the subchondral bone through the years and a concomitant increase in the stress level of the articular cartilage itself [11]. The effect is exacerbated by the fact that some of the protective mechanisms of energy absorption and joint congruity are lost as the bone stiffens. This means that the stress levels and amount of microfracturing will grow at an accelerating pace in a vicious circle effect, which, once begun, will be very difficult to reverse. It is this sequence of mechanical factors that we believe frequently culminates in age-related osteoarthrosis of the hip, knee, and other load-bearing joints [12].

It is not large mechanical insults that cause microfracture and eventually lead to osteoarthrosis; in the elderly, these cause instead such frank clinical fractures as Colles', and those to the hip and vertebrae. Normally, the impulsive forces generated by the events of daily life are attenuated by the controlled flexion (or extension) of joints, with the energy of the impacts being absorbed in muscle extension. Microfractures are the result of small, unexpected jolts for which the neuromuscular system is unprepared and cannot respond to in time. For example, if one unexpectedly steps off a curb that is four inches high, it takes only 144 milliseconds to hit the pavement, and this is short of the time it takes for a nervous response to travel from acceleration or pressure receptors to the brain and then to the muscles of the legs. Advancing age may well increase neuromuscular response times, so that the normal ability to absorb shocks deteriorates. Thus, the "clumsiness" of old age is probably another contributing factor to the high incidence of osteoarthrosis in the elderly. Also, the decline with age of the shock-absorbing ability of the intervertebral disks probably contributes to the almost universal radiographic evidence of osteoarthrosic change in the vertebral skeletons of the elderly [13].

In situations where stress concentrations in the joint become severe, the bounds within which Wolff's law operates may be exceeded, and bone formation may be blocked by excessive stress, while the resorption phase of remodeling goes on more or less normally. This imbalance can eventually produce a cyst [14]. Properly conceived and executed stress-relieving operations routinely lead to regression of such osteoarthritic cysts. When stress levels are reduced, the proper operation of Wolff's law is restored; the entire subchondral region returns to normal porosity and loses its sclerotic radiographic appearance [12, 15].

The formation of bony lips and spurs on the peripheral margins of a joint is also characteristic of osteoarthritis.

While the cause of these excrescences is not fully understood, they seem again to be a function of stress, since they only occur in load-bearing joints. They occur in the vicinity of capsular attachments and may represent remodeling responses stimulated by abnormally high forces in the capsule and associated ligaments of a mechanically distorted joint.

It should be emphasized that in early osteoarthrosis there is nothing abnormal about the bone tissue itself. Its appearance and structure may be altered by remodeling errors, but the calcified matrix is of normal composition and histologic organization. The skeleton maintains its ability to adapt and heal within a wide range of functional conditions.

OSTEOARTHROSIS AND OSTEOPOROSIS

Finally, in the context of the overall picture of age-related skeletal changes, it is interesting to note that senile osteoporosis tends to protect individuals against osteoarthrosis [16]. Since the development of the latter problem is predicated on the stiffening of subchondral bone, it is easy to see that this process is not likely to occur when the skeleton is experiencing a long-term net loss of bone mass. Instead of locally increased bone stiffness leading to cartilage erosion, one has a completely contrary result—the increasingly fragile trabeculae fracture, thereby absorbing energy and protecting the joints from damage [9]. It has been suggested that curing osteoporosis may make symptomatic osteoarthritis universal. Be that as it may, it is not now universal, and many people enjoy extended lifetimes without symptoms or evidence of articular cartilage loss. Clinical experience indicates that these individuals fall into two separate classes: those who are osteoporotic and those whose neuromuscular responses remain functional. This observation implies that it is important to bear in mind the intimate relationships between gerontologic problems. Solving the problem of osteoporosis may exacerbate the problem of osteoarthrosis, and the way to prevent that may be to solve the problem of neuromuscular degeneration.

REFERENCES

1. Sokoloff, L. *The Biology of Degenerative Joint Disease.* Chicago: University of Chicago Press, 1969.
2. Hill, A.V. Production and absorption of work by muscle. *Science* 131:897, 1960.
3. Yamada, H. *Strength of Biological Materials.* Baltimore: Williams and Wilkins, 1970.
4. Townsend, P.R., Raux, P., Rose, R.M., et al. The distribution and anisotropy of the stiffness of cancellous bone in the human patella. *J. Biomech.* 8:363–367, 1975.

5. Pugh, J.W., Radin, E.L., Rose, R.M. Quantitative studies of human subchondral cancellous bone: its relationship to the state of its overlying cartilage. *J. Bone Joint Surg.* [Am] 56A:313, 1974.
6. Goodfellow, J., Bullough, P. The pattern of aging of the articular cartilage of the elbow joint. *J. Bone Joint Surg.* [Br] 49B:175, 1967.
7. Thaxter, T.H., Mann, R.A., Anderson, C.E. Degeneration of immobilized knee joints in rats. *J. Bone Joint Surg.* [Am] 47A:567, 1965.
8. Abernethy, P.J., Townsend, P.R., Rose, R.M., Radin, E.L. Is chondromalacia patellae a separate clinical entity? *J. Bone Joint Surg.* [Br] 60B:205–210, 1978.
9. Byers, P.D., Contepomi, D.A., Farkas, T.A. A post-mortem study of the hip joint. *Ann. Rheum. Dis.* 29:15–31, 1970.
10. Meachim, G. The effect of scarification on articular cartilage in the rabbit. *J. Bone Joint Surg.* [Br] 45B:150, 1963.
11. Radin, E.L., Parker, H.G., Pugh, J.W., et al. Response of joints to impact loading. III. Relationship between trabecular microfractures and cartilage degeneration. *J. Biomech.* 6:245, 1973.
12. Radin, E.L., Paul, I.L., Rose, R.M. Current concepts of the etiology of idiopathic osteoarthrosis. *Bull. Hosp. Joint Dis.* 38:117–119, 1977.
13. Sylven, B. On the biology of nucleus pulposus. *Acta Orthop. Scand.* 20:275, 1951.
14. Freeman, M.A.R. Pathogenesis of osteoarthrosis: a hypothesis. In Apley, A.G. (ed.), *Modern Trends in Orthopedics*. London: Butterworth, 1972, p. 40.
15. Maquet, P., Radin, E.L. Osteotomy as an alternative to total hip replacement in young adults. *Clin. Orthop.* 123:138, 1977.
16. Foss, M.V.L., Byers, P.D. Bone density, osteoarthrosis of the hip, and fracture of the upper end of the femur. *Ann. Rheum. Dis.* 31:259–264, 1972.

Total Hip Replacement in the Elderly

▶

Carl L. Nelson, M.D.

15

Total joint replacement arthroplasty became a reality when John Charnley performed the first cemented total hip replacement approximately 20 years ago. Since then, total joint replacement arthroplasty has become a common orthopedic procedure. A consensus development conference was held in 1982 by the National Institutes of Health to determine the future and present status of total hip replacement arthroplasty in the United States. They estimated that 75,000 total hip replacements alone are performed on 65,000 patients annually in the United States. Of these, 60 percent are over the age of 65. The most common reasons for total hip replacement arthroplasty are osteoarthritis, 60 percent; fracture/dislocation, 11 percent; rheumatoid arthritis, 7 percent; aseptic necrosis, 7 percent; and revisions of previous hip operations [1].

PATIENT POPULATION

Candidates who are chosen for joint replacement surgery are generally those who have severe pain about the hip that is disabling, and the majority of the patients, of course, are elderly. In general, the total hip replacements that are used in the United States have three basic similarities: (a) they require acrylic bone cement to seat the prosthesis, (b) the acetabular component is made of high-density polyethylene, and (c) the femoral component is made of a biocompatible alloy. The two major considerations in surgery are the type of approach to be used and whether or not the greater trochanter is to be removed. Those who advocate one approach over another have shown no really hard data that advocate and prove which approach is superior, and it would appear in general that it is the surgeon's preference. In general, most surgeons do not remove the greater trochanter unless they feel it is necessary. It would appear from a scientific point of view that the trochanter may or may not be removed, but if it is necessary for visualization, it should be removed.

RESULTS

The end results of total hip replacement are indeed good. Approximately 90 percent of the patients who have had no previous hip surgery and have a total hip replacement arthroplasty should achieve a hip that is essentially pain-free, and the patient should be able to walk without limp and have a good range of motion. These results should be maintained for at least 5 years and possibly 20 years. Total hip replacement for the elderly, then, represents a dramatic improvement in the care of these patients that produces an exacting percentage of good and lasting results.

PROBLEMS AS RELATED TO THE ELDERLY

Joint Sepsis

One of the most disabling infection complications of joint replacement surgery is joint sepsis. The startling morbidity and, in some series, the suicide rate is as high as 10 percent following the complication of sepsis. There is little doubt that prevention rather than treatment of the sepsis is the key. The development of the clean-air room was initially popularized by John Charnley because of his initial high infection rate. Unfortunately, the clean-air room (laminar air flow) caused bitter debate among leading medical experts. Although the concept that bacteria should not be deposited into an open wound has been known since the time of Lister, there has been a resistance to apply the principle further to surgery for various vague arguments that have never been clearly elucidated. It has been well confirmed that clean-air systems that combined a laminar flow unit with a surgical isolator for the operating room crew markedly reduced the bacterial contamination of the wound.

Current studies have shown that clean-air systems are effective in reducing contamination, and a study from England [2] shows further that clean rooms combined with the use of hoods and preventive antibiotics produce the lowest rate of sepsis. At present, many physicians have not accepted either the concept of hoods or the clean-air system. In the future, there will be little to recommend standard operating theaters for total joint replacement surgery, and indeed clean-air systems with isolation of the surgeon would seem to be the preferred method of modern treatment.

Preventive antibiotics are also used, and although for some time there have been few hard data on which everyone could universally agree, the consensus has been that preventive antibiotics are effective. New and more recent studies have shown that antibiotics used preventively for a short duration are indeed effective and are used almost universally. Antibiotics in acrylic cement used as a prophylactic agent have not met with complete acceptance in the United States but have been used extensively in Europe. There are no reliable supportive data that unequivocally prove that these agents are effective in preventing infection.

Thromboembolic Diseases
Thrombophlebitis and pulmonary embolism are still significant problems; however, preventive therapy is well accepted and used throughout the United States. These therapies vary with the surgeon's preferences, and although the problem is dealt with, there is still a great deal to be learned and proved. Virchow's initial concept was that a state of hypercoagulability, venostasis, and injury to the vein are essential to the development of thrombosis. Little hard information has been added since that time. We have identified the high-risk patients in orthopedic surgery, who are those undergoing total joint replacement surgery of the lower extremity. Without prophylaxis the incidence of thrombophlebitis is high and the rate of fatal pulmonary embolism is unacceptable. While the various modalities that are used to protect the patient appear to be effective, we are much less effective in identifying the precise patient who will have significant pulmonary embolism. Therefore, further studies must be done to look for a more exacting method of identification and treatment. The problem is not unusual, since it is estimated that more than 150,000 to 200,000 fatal pulmonary emboli occur each year, of which 25 to 50 percent may be preventable.

Loosening
Loosening is a problem that usually affects the "younger elderly." The patient who is operated on after the age of 65 and who has modest expectations can probably expect a good result for a significant time. The younger patient, or the younger elderly if you will, who is large and has a very active life-style may develop the problem of loosening. At present, orthopedic surgeons have strived to improve the cementing techniques, which should produce a more lasting surgical result. The improvements in surgical technique essentially

make the 1980s' surgical technique for total hip replacement arthroplasty far more refined and markedly different than that performed in the early 1970s.

A newer development, which has been utilized in younger patients, is the porous-coated implant. These implants rely on bony ingrowth into the metallic surface, which is porous. This is a developmental procedure, but it is designed to do away with the need for bone cement and will improve bone fixation by the use of the patient's natural ability to heal and grow bone into the apertures in the metal. There are three major difficulties with porous-coated implants. First, the implant must be designed so that it achieves bone stability, which currently can only be effected through exacting surgical technique. If not, macro- or micromovement impedes or stops the ability of bone to grow into the apertures in the porous coat. Second, there is some question about the increased surface area and the absorption of ions. Last, once seated, the prosthesis may be difficult to remove. At present the techniques are developing for the hip, but the most data available for scrutiny are for the knee, in which porous-coated implants have been used widely. The early results appear encouraging.

Tumors in Joint Replacement Surgery
The use of joint replacements for the treatment of tumors of the hip is developing and is an area in which patients with large tumors about the hip can be treated by custom-constructed prostheses. This development adds a significant advantage to the patient who would be bedridden and now can be mobile because of the use of the acrylic bone cement custom-designed for large replacement prostheses and surgical techniques.

Specially Designed Prostheses
At the present time, some centers have utilized computerized designed prostheses for complex patient problems. This technology is helpful in revision surgery where special prostheses with specific modifications are needed. Our experience has been more than satisfactory when special prostheses are needed; the x-ray tracings are made, the specific recommendations are sent to the manufacturer, who places them on a computer, which then sketches out and designs the prosthesis specifically for the individual.

Nutritional Status
One of the developments that has been further pursued in joint replacement surgery and has significantly affected the elder patient is the evaluation of the patient's nutritional

status prior to surgery. A recent study pointed out that at least 40 percent of elderly patients appearing for joint replacement surgery have poor nutritional status [3]. Correction of this problem can lead to improvement in the patient's risk for postoperative morbidity and mortality.

PROSPECTUS

In the future, patients will be evaluated more carefully prior to surgery by studying both their nutritional status and their ability to combat infection through other host-mediated responses. Surgery will be performed in an ultra-clean room that essentially ensures that no bacteria will contaminate the patient; anesthetic will be formulated specifically to reduce the incidence of thrombophlebitis and provide a dry field; and the surgery will be done with the most technologically advanced instrumentation. Postoperatively, the rehabilitation phase will be active and early, and the patient can look forward to a good long-term result.

With time, various methods of prosthesis fixation will substitute for acrylic bone cement, and the long-term wearability problems of polyethylene will be solved by use of new designs. Orthopedics in the field of joint replacement for the elderly is indeed in the forefront of modern technology and has represented a major improvement in patient care.

CONCLUSIONS

The results of total hip replacement arthroplasty are excellent, and total joint replacement arthroplasty has been a significant advance in modern medicine.

1. "Younger elderly" patients who develop loosening still pose a problem, but improved techniques, better materials, computerized design technology, and porous implants seem to be alleviating this problem.
2. The wearability of polyethylene is not a problem at present but may lead to problems in the future.
3. Although thrombophlebitis and pulmonary embolism remain significant clinical problems, treatment modalities for these complications are well accepted and utilized.
4. Sepsis is a continuing problem. However, with improvements in the patient's nutritional status, clean-air systems, and so on, the rate of infection can be reduced further.

Although this chapter discusses total hip replacement arthroplasty, the principles, complications, and difficulties are indeed very similar for all joint replacement surgery for the elderly.

REFERENCES

1. National Institutes of Health Consensus Development Conference. Total Hip Joint Replacement, Bethesda, Md. March 1–3, 1982.
2. Lidwell, O. Investigation of the effect of ultra-clean air in operating rooms on surgical sepsis. Paper presented at the American Academy of Orthopaedic Surgeons, New Orleans, January 1982.
3. Jensen, J.E., Jessen, T.G., Smith, T.K, et al. Nutrition in orthopaedic surgery. *J. Bone Joint Surgery* [*Am.*] 64A:1263–1272, 1982.

Trauma in the Elderly

▶

Gordon A. Hunter, M.B.
E.T.R. James, M.B.

16

No one is so old as to think he cannot live one more year.
—Cicero, 100 B.C.

Trauma can be defined as a dynamic transfer of energy from one system in a higher state of energy to a physiologic system in a lower state of energy. For the purposes of this chapter, the elderly patient will be assumed to be over the age of 65 years.

Lawton [1] states that of approximately 16 million elderly people aged 65 and over in the United States, 25,000 die and 3 million are injured by accidents each year. This age group constitutes less than 10 percent of the population, but 25 percent of accidental fatalities [2].

Three types of accidents account for the majority of accidental deaths in elderly people: falls, burns and fires, and motor vehicle accidents.

FALLS

Falls are the largest single cause of accidental death in the elderly and account for more than half of all deaths of persons over the age of 74 years [3]. Many more are disabled by falls in the home, hospital, nursing homes, and community. In 1974 in Canada, more than 1800 deaths were due to accidental falls; 72 percent of these deaths were in persons over 65 years of age [4]. Some of the causes of falls in the elderly are listed in Table 16-1.

Religious, civic, and private institutions nowadays have taken over the care of many elderly patients previously cared for by the middle-aged family unit. In these nursing homes there may be minimal staffing and superficial medical care of confused, nonproductive persons; most accidents occur at night, the time of least staff assignment.

Improved architectural features should be incorporated in private and nursing homes to reduce the incidence of falls in the elderly (e.g., floor-level lighting to compensate for poor vision, loss of postural balance), and careful attention should be paid to flooring, stairs, and furnishings. There must be changes in design of wheelchairs, beds, chairs, and bathroom space.

Table 16-1. Causes of Falls in the Elderly

Cardiovascular	*Musculoskeletal*
Drop attacks	Arthritis
Postural hypotension	Fractures
Myocardial ischemia	Bone disease
Dysrhythmias	Neuromyopathy
Cerebrovascular	*Other factors*
Postural imbalance	Urinary problems
Transient ischemic attacks	Mental confusion
Giddiness	Loss of vision
Hemiplegia	Effect of drugs or alcohol
Epilepsy	Preexisting disease
	Environmental factors

BURNS AND FIRES

Bull [5], analyzing a large series of burns at the Birmingham Accident Hospital, was able to show that patients over the age of 65 would be expected to have a 50-percent mortality after only 17 percent of the body surface was burned, compared to 56 percent of the body in young adults between the ages of 15 and 44 years.

To reduce the incidence of accidental burns and fires, all nursing-home personnel should be trained in fire prevention, and regular fire drills should be obligatory. Fires from cigarette butts and defective electric wiring account for a large number of such accidents. The gas jet on the stove may be turned on and forgotten, and consideration should be given to the use of a gas with a more penetrating odor. Dials on stoves should be distinctly marked so that the on-off positions can be easily seen or felt. Nonflammable cotton should be used in pajamas and nightgowns and should be legislated for use for the elderly, as it has been for young children. Burns in the bathtub are quite frequent; preset temperature controls and the use of a seat in the tub would help to prevent such accidents.

MOTOR VEHICLE ACCIDENTS

Baker and Spitz [6] reported that a decreased ability to survive crashes caused older persons to be greatly overrepresented among fatally injured drivers. Baker et al. [7] introduced an abbreviated injury scale (A.I.S.) as a method for describing patients with multiple injuries and evaluating emergency care. The results of their study confirmed that elderly patients have a poor prognosis following multiple injuries, but use of the A.I.S. made it possible to determine that this increased mortality in the elderly is most pronounced when the injuries are least severe.

Bull [8], using the injury severity score (I.S.S.), confirmed that there was an expected mortality of 50 percent (lethal dose, or LD) when the I.S.S. was only 20 in a patient over the age of 65 years, compared with an I.S.S. of 40 in a patient under the age of 44 years.

We reviewed our own experience at Sunnybrook Medical Centre Regional Trauma Unit in Toronto, Ontario. During a 3-year period, from 1976 to 1979, there were 390 patients admitted with multiple trauma; of this group, 19 patients were aged 70 years and over, and these patients provide the basis for this review. The majority of admissions were due to pedestrian accidents and falls. There was a high incidence of respiratory complications from rib fractures; the risk of chest infection from prolonged intubation (endotracheal and intercostal) is especially high in the elderly patient. Our results, showing a high mortality with increasing age and a lower I.S.S., confirmed the findings of Baker et al. [7] and Bull [8]. There were 11 survivors and 8 deaths. The major causes of death in elderly patients were head and spinal cord injuries and respiratory and cardiac failure. More than 60 percent of the deaths ocurred in the first week after trauma. Of the patients who died, all had multiple injuries and had an LD of more than 50.

To reduce trauma in the elderly, one should consider the use of brightly colored clothing for pedestrians at night, and the elderly driver should avoid driving in bad weather and at night. Mandatory use of seat belts should be legislated and enforced. Civic authorities should pay particular attention to lighting at intersections, build graded underpasses on wide roads, and facilitate both entry and exit areas in stores and on public transportation.

CONCLUSIONS

In patients over 65 years of age, accidents are as important a cause of death as pneumonia and diabetes. The elderly are more liable to sustain an accident than the young and are more liable to suffer permanent disability and increased morbidity and mortality from trauma. Commonly overlooked musculoskeletal problems leading to severe morbidity and death in the elderly include

1. Fracture of the ribs and pelvis, leading to unrecognized blood loss
2. Head injuries, resulting in subdural hematoma
3. Fractures of the cervical spine, which cause serious spinal cord problems

To reduce the incidence and severity of complications after trauma in the elderly, the anesthetist should remember the greater sensitivity of the elderly to preoperative medication and the importance of restoring fluid and electrolyte balance. The anesthetist should also be sure to avoid hypoxia, hypotension, and hypothermia in the operating room. All trauma patients must be considered as having full stomachs and therefore to be at risk of aspiration; use of a nasogastric tube and strict attention to the airway are essential. Rapid transfusion for blood and fluid loss through two large-bore intravenous cannulae should be combined with the use of a Foley catheter to monitor urine output, and invasive arterial pressure monitoring may be combined with a Swan-Ganz catheter to monitor cardiovascular dynamics.

The surgeon should attempt to combine speed with safety and use prophylactic antibiotics and anticoagulants to reduce postoperative complications. Early fixation of fractures, by internal or external devices, reduces pain and the incidence of posttraumatic pulmonary complications and permits improved wound management of open fractures.

The physiotherapist should encourage early motion of all injured joints and pay particular attention to postoperative pulmonary complications such as lobar collapse and aspiration pneumonia. Early ambulation will avoid prolonged hospitalization and enable the elderly patient to return to the home or nursing home as soon as possible.

Table 16-2 lists some of the particular reasons for the increased morbidity and mortality following trauma in the elderly.

If the early management of the severely injured elderly patient is not successful and posttraumatic problems such as multiple organ failure develop, one should not persist in fruitless diagnostic and therapeutic procedures. One should allow the elderly patient to die under such circumstances in relative comfort and dignity. One should remember the words of W. Somerset Maugham, who, at the age of 90 years, said, "I am sick of this way of life. The weariness and sadness of old age make it intolerable. I have walked with death in hand, and death's own hand is warmer than my own. I don't wish to live any longer."*

Table 16-2. Reasons for Increased Morbidity and Mortality in the Elderly Following Trauma

Reduced life expectancy
Preexisting acute or chronic illness
Accidental hypothermia
Delay in medical attention
Increased incidence of pulmonary infection, thromboembolism, myocardial and cerebral infarction after surgery
Poor response to fluid and blood loss
Increased sensitivity to drugs or alcohol
Prolonged hospitalization increases the risk of hospital-acquired infection

REFERENCES

1. Lawton, A.H. Accidental injuries to the aged. *Gerontologist* 5:96–100, 1965.
2. Waller, J.A. Injury in aged: clinical and epidemiological implications. *N.Y. State J. Med.* 2200–2208, 1974.
3. Rodstein, M. Accidents among the aged: incidence, causes and prevention. *J. Chronic Dis.* 17:515–526, 1964.
4. Statistics Canada. *Canada's Elderly.* 1976 Census of Canada (Ottawa).
5. Bull, J.P. Revised analysis of mortality due to burns. *Lancet* 2:1133–1134, 1971.
6. Baker, S.P., Spitz, W.U. Age effects and autopsy: evidence of disease in fatally injured drivers. *J.A.M.A.* 214:1079–1088, 1970.
7. Baker, S.P., O'Neill, B., Haddon, W., Jr., Long, W.B. The injury severity score: a method for describing patients with multiple injuries and evaluating emergency care. *J. Trauma* 14:187–196, 1974.
8. Bull, J.P. The injury severity score of road traffic casualties in relation to mortality, time of death, hospital treatment time, and disability. *Accid. Anal. Prev.* 7:249–255, 1975.

*In M.B. Strauss (ed.), *Familiar Medical Quotations.* Boston: Little, Brown, 1968.

Hip Fractures in the Elderly

▶

Anthony P. Dwyer, M.D.
Carl L. Nelson, M.D.

17

The high morbidity and mortality associated with hip fractures in the elderly are clear examples that all the aging and deteriorating body systems cannot cope with the added stresses of the immobilization, anesthesia, surgical trauma, and loss of function.

MORTALITY

The orthopedic approach to the problems of hip fractures has varied from achieving better reduction and more rigid fixation of intertrochanteric fractures or replacement of femoral heads doomed to avascular necrosis following femoral neck fracture to recommending nonoperative treatment [1-3]. Despite this wide spectrum of treatment regimens and the improvements in patient care, studies show that both the hospital and 3-month mortality rates have remained unchanged [4, 5].

Hip fractures have therefore been viewed pessimistically since 1935, when Speed [6] named the femoral neck fracture "the forgotten fracture"—a term that Hunter [7] more recently applied to intertrochanteric fractures. Mortality associated with hip fractures increases according to the length of time following the fracture [5]. Mortality is also related to the age and sex of the patient, the associated medical diseases, the severity of the fracture, and the method of treatment [5]. These variables can account for the differences in reported mortality rates and need further evaluation before the causes of the high mortality can be elucidated.

Miller [8], in a retrospective review of 360 patients, found a mortality of 27 percent 1-year postfracture, as compared with 9 percent for an age-matched population. This increased mortality was marked over the first 4 months and presented for the first 8 months postfracture. Miller found that the only significant factors were being over 80 years of age, having cerebral dysfunction at the time of fracture, and being male. He could not establish any significant relationship between the duration of the anesthetic, the blood loss, the fixation device, or the number of postoperative complications and the outcome at 1 year. However, there was an increased incidence of pulmonary disease, diabetes, and cerebral dysfunction in this group of patients. Miller stated that the total number of hospital days related to hip fractures numbered tenth overall. Lewinnek et al. [3] estimated that deaths following hip fractures make up 2 percent of the total annual U.S. death rate.

Sherk et al. [9] reported an improved 4-month mortality of 28 percent in psychiatric patients receiving prompt surgical treatment. Sherk et al. [10] had previously reported a pessimistic 4-month mortality of 54 percent following operative and 64 percent following nonoperative treatment.

Jensen and Tondevold [5], reporting on 1,592 hip fracture patients, found a hospital stay of 24 days, a 3-month mortality of 17 percent, and a 6-month mortality of 21.5 percent. It took 1.6 years before the survival rate following hip fracture paralleled the survival rate for the age-matched general population.

Dahl [4], however, reviewing 675 patients, found a higher hospital mortality rate of 13.9 percent. *The mortality during the first month postfracture was approximately 15 times the expected death rate, and the mortality during the second month was approximately 7 times the expected death rate of the age-matched general population. Those surviving the second month had no added risk of dying during the following 4 years.*

Dahl's study [4] showed that the mortality following hip fracture was related to patient age and sex, as patients over 75 years old and men had a higher death rate. Laskin et al. [11] reported a 6-month mortality of only 7 percent in 236 patients healed under a strict protocol of adequate medical management of cardiac, pulmonary, and skin problems; rapid rehydration; and rigid fixation of the intertrochanteric fracture followed by early mobilization and immediate weight bearing.

On the other hand, Lyons and Nevins [12] pessimistically recommended nonoperative treatment of senile patients but did so after having reviewed only nine patients. These authors consider it unrealistic to place the blame for these high mortality rates on purely orthopedic matters without considering the prefracture medical and functional status of the patient. As Dahl [4] has clearly pointed out, the severe associated disorders that were particularly significant preoperatively were recent myocardial infarction, renal failure, uncompensated congestive cardiac failure, and cancer. More than 75 percent of the hospital deaths were due to bronchial pneumonia, congestive cardiac failure, and cerebral vascular accidents, while pulmonary embolism accounted for only 8 percent of hospital mortality.

Summary

1. The average 6-month postfracture mortality rate is approximately 20 percent.
2. Mortality is markedly increased during the first 4 months and returns to that of the general population after 1 year.
3. Mortality in patients under 60 years of age is not significantly higher than normal.
4. Patients of advanced physiologic age (over 80 years) had markedly increased mortality.
5. Mortality can be decreased with expert medical preoperative preparation, prophylactic antibodies, prompt expert operative treatment, early mobilization, and immediate weight bearing.

MORBIDITY

The morbidity associated with hip fractures has not received the same emphasis as the mortality. However, any complication that tends to prolong the hospital stay and to delay the return of the patient to his former social situation is both significant and costly. Pain and other complications that reduce the patient's mobility and walking function will limit rehabilitation and reduce the quality of life.

Femoral Neck Fractures

Hunter [2] gives a pessimistic review of the complications of primary prosthetic replacement, reporting a dislocation rate of 11 percent; deep wound infection, 8 percent; bronchial pneumonia, 18 percent; thromboembolism, 10 percent; cerebrovascular accident, 3 percent; myocardial infarction, 1 percent; and pressure sores, 4 percent. Other authors have quoted a dislocation rate of between 0.3 and 10 percent and an infection rate varying between 2 and 42 percent. Fielding [13] reported a union rate of 90 percent using the Pugh nail and an avascular necrosis rate of 11 percent for undisplaced fractures and 23 percent in the united displaced fractures.

Other workers have reported on a union rate varying between 60 and 100 percent and an avascular necrosis and late segmental collapse rate between 17 and 84 percent [13a]. These rates vary with the degree of treatment and the method used to diagnose the avascular necrosis.

Intertrochanteric Fractures
Laros and Moore [14], in a retrospective x-ray review of surgically treated intertrochanteric fractures, concentrated on the effects of complications of fixation (as defined by Dimon and Hughston [15]), classifying the fractures as stable or unstable and the reduction as anatomic or nonanatomic. They found that complications of fixation occurred at a significantly greater rate in varus as compared with valgus reductions. They also found that there was a decreased rate of complication of fixation in most fractures with a Singh's [16] osteoporosis grading of 4 and above in both stable and unstable fractures. There was an increased incidence of hip pain (particularly after nail penetration) and repeat surgery in the *complication of fixation* group.

Complications of fixation did not significantly decrease walking capacity, which was related to length of follow-up. The authors reported a nonunion rate of 6 percent in the unstable fractures and 1 percent in the stable fractures.

Contrary to Sarmiento [17], who placed great emphasis on the need for cortical alignment to avoid instability, Laros and Moore [14] found no significant difference in the complication rate between anatomic and nonanatomic reduction in both the unstable and stable fracture groups. They point out that 61 percent of the patients with *complications of fixation* had no pain and 71 percent of these complicated patients did not require any secondary surgery. As complications of fixation did not significantly affect the ambulatory status, it appears that mobilization was achieved in spite of these fixation complications.

REHABILITATION

Laros and Moore [14] state, "the primary goal of hip fracture surgery in the elderly is fracture stabilization permitting mobilization of the patient and avoidance of complications of bed rest." For Ceder et al. [18–20] and Barnes et al. [21], the principal goal for the elderly with hip fractures is to return them home. When discussing rehabilitation, Ceder et al. [19] reports "with the technical improvements of the last decade that hip fractures per se are of less importance for recovery than sociomedical factors." Ceder et al. [20] found three important prognostic factors, the most important being the *patient's health*, while the others were whether the *patient was living with someone* and *the patient's ability to talk*. He found that age was of less importance than the general medical conditions, with neurologic disorders being the greatest threat. If all three of his important prognostic factors were positive, 90 percent of his patients returned home, compared to only 50 percent when only one of the prognostic factors was present.

Ceder et al. [19] classified the patients according to their general medical condition:

- Group A: No known disease or impairment other than the hip fracture
- Group B: Additional disease not likely to impede rehabilitation
- Group C: Additional disease that will potentially affect rehabilitation

Ceder et al. [19] found that only half of the group C patients were discharged directly home. He found that the main recovery in walking, activities of daily living, and household activities took place within the first 4 months, and only the ability to go shopping improved after this time.

Ceder et al. [19] sum up, "The rehabilitation program may not affect the mortality but it can enable more to live as normal a life at home for the period they do survive."

Cobey et al. [22] found that the best indicators of recovery following hip fractures were

1. Physiotherapy assessment postoperatively
2. Rate of "out of house" activity prior to fracture
3. Patient's mental clarity

These factors were the important and significant factors, while the severity of the fracture showed no significant relationship to the outcome.

At 4 months postoperatively the significant factors were

1. Two-week postoperative walking ability
2. Patient's age
3. Rate of visiting before the injury

Miller [8] found that the ambulatory status (at 1 year postfracture) was significantly related to advanced physiologic age and cerebral dysfunction.

TREATMENT RECOMMENDATIONS

The functional goals of treatment vary according to the prefracture activity of the patient. Halpin and Nelson [23] classify the patients as

1. Active
2. Semiactive
3. Bedridden

Intracapsular Hip Fractures

Intracapsular hip fractures are classified using the Garden classification [21, 24].

Garden Types I and II In Garden types I and II, multiple parallel pin fixation is recommended. This can be performed with a percutaneous technique, which results in a low infection rate and low operative blood loss and requires a minimum of anesthesia. This technique virtually ensures that these fractures will not become displaced.

Garden Type III Garden type III in a *fully active* patient with good bone quality should receive a closed reduction and multiple parallel pin fixation.

For the *semiactive* or *bedridden* patient or any patient with poor bone quality, we recommend endoprosthetic replacement.

Garden Type IV In Garden type IV, we recommend Endoprosthetic replacement for *bedridden* and *semiactive* patients. In the *fully active,* vigorous patient, the recommendation is for an open anatomic reduction, rigid internal fixation. The addition of a posterior muscle pedicle graft of the Meyers variety [25] may be indicated.

Extracapsular Hip Fractures

The extracapsular or intertrochanteric fractures are classified according to Ender and Simon-Weidner [26]. They classified two major types of intertrochanteric fractures—the "gaping" and the "impacted" variety.

Gaping Intertrochanteric Fractures The gaping intertrochanteric fractures are those that seem to open inferiorly like a book. The degree of severity is reflected by the degree of comminution and the soft-tissue disruption as seen on x-ray. Gaping fractures are further subdivided into "simple," "posterior wedge," and "lateral displacement" varieties. The simple variety is not comminuted and has a low incidence of complication. The posterior wedge variety is less stable after fixation. The lateral displacement variety has elevation of the femoral shaft fragment and is the least stable because of the degree of soft-tissue disruption.

Impacted Intertrochanteric Fractures The impacted variety is classified according to the "spike"—i.e., the shape—of the proximal fragment. The spike shape of the proximal fragments renders fracture reduction and nailing more difficult.

Treatment The generally accepted method of treatment of intertrochanteric fractures is closed reduction and fixation with a sliding or dynamic compression hip screw [11]. This method of treatment reduces the incidence of malunion (usually varus or external rotation) and nonunion. However, the method is associated with significant wound infection and blood loss.

The condylocephalic nailing of Ender is specifically indicated in institutionalized elderly patients with cerebral dysfunction and in patients who have a debilitating general medical condition. These patients are poor candidates for standard surgical and anesthetic procedures and have poor rehabilitation prognosis. Reviews by Ender and Simon-Weidner [25] and Nelson et al. [27] report a lower infection rate, less operative blood loss, and reduced general morbidity after condylocephalic nailing.

The recommended treatment of intertrochanteric fractures in the vigorous, *active patient* with good bone stock is anatomic reduction and a sliding or dynamic compression hip screw, as these patients tolerate the increased operative blood loss. The increased risk of infection is acceptable, while any degree of malunion in varus or external rotation is not. For *bedridden*, institutionalized patients, the goals of treatment are pain relief, prevention of long immobilization, and improved nursing facility. These goals are adequately achieved with the use of the Ender nail, particularly in experienced hands, without the risks of infection and increased blood loss. *Semiactive* patients can be treated either with anatomic reduction and a compression fixation screw or with the Ender nail, depending on the medical status and individual needs of the patient and the surgical skill and experience of the surgeon.

CONCLUSIONS

Thinking individuals would recommend a more organized approach to treatment of fractures about the hip. In an idealized future, the elderly patient would be completely evaluated by a team of specialists (orthopedists, internists, anesthesiologists, and nutritionists) and would be stabilized physiologically.

An anesthetic with minimal side effects would be given by individuals dedicated to and familiar with the problems of the hip and the aged, so that the operation would be done with minimal time and blood loss and maximal benefit, since the procedure would be tailored to the individual patient on the basis of his or her need rather than those of the surgeon's usual procedure. The surgery would be done by the most experienced and talented surgeon available, especially for the very ill patient.

Following surgery, appropriate nutrition, rehabilitation,

and psychological treatment would make up the third and equally important phase of patient care so that the patient can return to his or her life.

We must ask why the results of treatment of hip fractures in the elderly are so unacceptable and so inferior to those of total hip replacement, which requires more extensive and difficult surgery. We believe the viable goal for the future is to establish specialized units and centers for fractures of the hip to reduce the unacceptable morbidity and mortality so that these fracture patients will no longer be "forgotten."

REFERENCES

1. Evans, E.M. The treatment of intertrochanteric fractures. *J. Bone Joint Surg.* [*Br.*] 33B:192, 1951.
2. Hunter, G.A. Should we abandon primary prosthetic replacement for fresh displaced fractures of the neck of the femur? *Clin. Orthop.* 152:158, 1980.
3. Lewinnek, G.D., Kelse, J., White, A.A., Kreiger, N.G. The significance and a comparative analysis of the epidemiology of hip fractures. *Clin. Orthop.* 152:35, 1980.
4. Dahl, E. Mortality and life expectancy after hip fracture. *Acta Orthop. Scand.* 51:163, 1980.
5. Jensen, J.S., Tondevold, E. Mortality after hip fractures. *Acta Orthop. Scand.* 50:161, 1979.
6. Speed, K. The unsolved fracture. *Surg. Gynecol. Obstet.* 60:341, 1935.
7. Hunter, G.A. The results of operative treatment of intertrochanteric fractures of the femur. *Injury* 6:202, 1974.
8. Miller, C.W. Survival and ambulation following hip fracture. *J. Bone Joint Surg.* [*Am.*] 60A:930–934, 1978.
9. Sherk, H.H., Snape, W.J., Lopite, F.L. Internal fixation versus nontreatment of hip fracture in senile patiens. *Clin. Orthop.* 141:196, 1979.
10. Sherk, H.H., Crouse, F.R., Probst, C. The treatment of hip fractures in institutionalized patients: comparison of operation and nonoperative methods. *Orthop. Clin. North Am.* 5:543, 1974.
11. Laskin, R.S., Gruber, M.A., Zimmerman, A.J. Intertrochanteric fractures of the hip in the elderly: a retrospective analysis of 236 cases. *Clin. Orthop.* 141:108, 1979.
12. Lyons, J.L., Nevins, M.A. Nontreatment of hip fractures in senile patients. *J.A.M.A.* 238:1175, 1970.
13. Fielding, J.W. The telescoping Pugh nail in the surgical management of the displaced intracapsular fracture of the femoral neck. *Clin. Orthop.* 152:123, 1980.
13a. Sevitt, S. Avascular necrosis and revascularization of the femoral head after intracapsular fracture. *J. Bone Joint Surg.* [*Br*] 46B:270, 1964.
14. Laros, G.S., Moore, J.F. Complications of fixation in intertrochanteric fractures. *Clin. Orthop.* 101:110, 1974.
15. Dimon, J.H., Hughston, J.C. Unstable intertrochanteric fractures of the elderly. *J. Bone Joint Surg.* [*Am.*] 49A:440, 1967.
16. Singh, M., Nagrath, A.R., Maini, P.S. Changes in the travecular pattern of the upper end of the femur as an index of osteoporosis. *J. Bone Joint Surg.* [*Am.*] 52A:457, 1970.

17. Sarmiento, A. Avoidance of complications of internal fixation of intertrochanteric fractures. *Clin. Orthop.* 53:47, 1967.
18. Ceder, L., Lindberg, L., Odberg, E. Differentiated care of hip fracture in the elderly: mean hospital days and results of rehabilitation. *Acta Orthop. Scand.* 51:157, 1980.
19. Ceder, L., Svensson, K., Thorngren, K.G. Statistical predication of rehabilitation in elderly patients with hip fractures. *Clin. Orthop.* 152:185, 1980.
20. Ceder, L., Thorngren, K.G., Wallden, B. Prognostic indicators and early home rehabilitation in elderly patients with hip fractures. *Clin. Orthop.* 152:173, 1980.
21. Barnes, R., Brown, J.T., Garden, R.S., Nicoll, E.A. Subcapital fractures of the femur: a prospective review. *J. Bone Joint Surg.* [Br] 58B:2, 1976.
22. Cobey, J.C., Kirby, J.H., Colant, L., et al. Indicators of recovery for fractures of the hip. *Clin. Orthop.* 117:258, 1976.
23. Halpin, P.J., Nelson, C.L. A system of classification of femoral neck fractures with special reference to choice of treatment. *Clin. Orthop.* 152:44, 1980.
24. Garden, R.S. Low angle fixation in fractures of the femoral neck. *J. Bone Joint Surg.* [Br.] 43B:647, 1961.
25. Meyers, M.H. The role of posterior bone grafts (muscle pedicle) in femoral neck fractures. *Clin. Orthop.* 152:143, 1980.
26. Ender, J., Simon-Weidner, R. Die Fixierung der trochenteren Brüche mit rundem elastichen Condylennägeln. *Acta Surg. Aust.* 1:40, 1970.
27. Nelson, C.L., Weber, M., Bergman, B., Gerdes, M. Ender nailing of intertrochanteric fractures. In Leach, R.E., Hoaglund, F.T., Riseborough, E.J., (eds.), *Controversies in Orthopaedic Surgery*. Philadelphia: W.B. Saunders, 1982, pp. 139–153.

Rehabilitation Medicine in Aging

▶

William J. Erdman II, M.D.

18

The word *rehabilitation* is derived from the Latin "rehabilitare," which was a process of restoring by formal act a person degraded or attended to former privileges, rank, possessions, and a person's good name or memory by authoritative pronouncement [1]. Only the king could restore an officer or a gentleman who had derogated from his rank.

In practice, rehabilitation is concerned with the restoration to maximum potential, from a physical, emotional, social, and vocational standpoint, individuals disabled in the neuro-musculo-skeletal systems. An aging person cannot be restored by action of a king, by the waving of a wand, or by the taking of any medicine. It requires a great deal of hard work and the interaction of many health professionals to retrain, strengthen, mobilize, encourage, and achieve reasonable goals. Frequently, the best that can be hoped for is changing the slope of the downhill course that the aging process normally imposes on the human body. But it is essential that dignity, physical function, and emotional integrity be maintained or restored as long as possible [2]. It is much easier for the family to cope with the problems of aging and to keep a relative out of an extended care facility when his sense of worth, as well as his physiologic integrity, is preserved.

THE REHABILITATION PROGRAM

Entering a rehabilitation program from an acute medical/surgical bed, from a difficult home situation, or from a nursing home environment may have a salutary influence on an individual who is depressed by the reality of his or her decaying physique and shrinking horizons [3]. The depression lifts and the patient begins to do things independently, requiring less assistance from attendants almost immediately. Frequently the effect is dramatic, occurring before any possible influence of the therapeutic milieu has had time to modify the individual physically. The fact that a decision has been made that the program is worth undertaking may significantly change the patient's direction of motivation from hopelessness to hopefulness. Everyone is motivated, but not always in a positive direction. A patient may be more willing to exercise and attempt activities of daily living for a professional team than he or she was at home or in a nursing home. When an individual is willing to participate and makes a conscious effort to cooperate, the principles of muscle reeducation, maintenance of range of motion, improvement in cardiorespiratory reserve, and progressive resistance exercises have relevance.

Even without active participation, there are advantages to physical activity in preserving bodily function by passive exercises. However, passive activity can never replace active exercise for increasing muscle strength and endurance or for positively influencing cardiovascular and respiratory function [4].

Some of the chemical, structural, strength, and physiologic changes proportional to advancing years are presented in this book. These changes affect and influence the rate at which exercises may be performed by an older person [5]. Joint stability and muscle strength may require additional external support (e.g., a walker, canes, crutches). Lightweight orthoses to stabilize weak joints and assist movement in a controlled manner are currently available [6]. They are cosmetically acceptable, easy to apply, and more consistent with physiologic requirements. Similarly, plastic prostheses that are lighter than the amputated body part and that weigh much less than the wood, leather, or metal prostheses of even a few years ago are being manufactured [7]. They are cosmetically more acceptable and are more easily applied than the older versions. Greater flexibility gives increased comfort.

Not all changes of aging are necessarily bad. Pain thresholds are frequently higher, and some destructive diseases progress less rapidly than in younger years.

Age is not the only determinant in a rehabilitation program, and many patients in their eighth or ninth decades respond well. All patients are placed on a regimen that is calculated and adjusted to the individual tolerances. The program includes medical management, rehabilitation nursing, occupational therapy, physical therapy, speech therapy, recreational therapy, and psychiatric support. After a thorough evaluation of the patient by each member of the team, the patient and his or her family participate in a conference at which they have the opportunity to exchange observations, project goals, and set an approximate time frame for their accomplishment. The patient and family are encouraged to ask questions and participate fully in the discussion and decision-making process. The only variation for a geriatric patient as compared with a younger one is that vocational goals are not included. The vocational history is important, frequently in predicting behavior in the program. Avocational interests are explored and supported. The program is reviewed, and if no hobbies or interests exist, efforts are made to establish some. The program is reviewed at least weekly by the team to adjust activities and goals.

Early in the process of assessing the rehabilitation goals for a patient, it is essential that the staff have as clear a perception of the home environment as possible. A visit by a staff member to assess the architectural and attitudinal barriers and gain a clear understanding of the matrix and constellation of the family will avoid the pitfalls, disappointments, and frustrations produced by unrealistic targets. Social workers trained to ferret out these problems skillfully contribute an essential ingredient to the evaluation process. Their input should be made at the beginning of and throughout the program, rather than at the discharge conference alone. This is even more important for elderly patients.

In order to assess the value of the program, each patient is reviewed at the time of discharge and on regular follow-up visits. The data are computerized so that there is a continuous update of the efficacy of the rehabilitation program. The information helps to make adjustments, evaluate cost-effectiveness, and more importantly for the individual patient, determine if he needs additional services. The patient's condition and abilities on admission, on discharge, and at follow-up are measured and recorded [8].

GOALS

The primary goal of many aging patients is to resume ambulation. Any activity that is not directly associated with walking does not interest them. Frequently, they need to be convinced about the value of occupational and speech therapy sessions. These may be particularly stressful because they focus on areas of disability that may not recover as well as ambulation. In fact, those individuals with cognitive disorders may be totally unaware of significant deficits, which are brought to the forefront in occupational and speech therapy sessions. It is not the therapies that are disagreeable but rather the threat that is imposed by having to recognize the losses and the sometimes painful efforts to learn substitution and compensation.

The greatest problem for families accepting responsibility for taking care of a member is that of incontinence, especially fecal. This issue is the primary determinant in placement. Other deficits, such as dressing, ambulation, speech, loss of motion, inability in self-care, are easier to accept. The need for constant, regular, or even frequent cleaning up and the attendant odors is the single most frequent divisive factor in choosing between living with the family and in an extended care facility. At home the responsibility falls most heavily on a reluctant daughter or daughter-in-law. The frequency of incontinence can be significantly decreased by bowel and bladder training, timing, physical activity, and medications.

In physical therapy or in the exercise aspects of nursing care and occupational and recreational therapy, those principles applicable to a younger person are also operative. Increasing strength requires working muscles against resistance, whether individual, mechanical, or gravitational. Increasing resistance is necessary to increase muscle strength. Restoration or maintenance of endurance depends on increasing the frequency of repetitions.

Sometimes it is unwise to achieve greater independence of support. It is much better to use a cane, walker, or human assistance than to achieve independence, only to invite new fractures and bruises that can complicate the underlying problems.

CONCLUSIONS

Much has been written about increasing the life span in this century from the mid-forties to mid-seventies. The percentage of the population in the seventh, eighth, and ninth decades and beyond shifts to the right constantly. In fact, the fastest increasing cohort is the population in the eighth decade. Medical progress, eradication of disease, better sanitation and public health measures, and improved nutritional knowledge are but some factors contributing to change. The staggering social and economic impact of these changes causes increasing concern to planners in government and society. Rehabilitation is the logical antidote to passive or active euthanasia. George Morris Piersol expressed the challenge beautifully when he enunciated, "We have added years to the lifespan and we must add life to years."

REFERENCES

1. *Oxford University Dictionary on Historical Principles*, 3rd ed. Oxford, England: Clarendon Press, 1955.
2. Fordice, W.E. Psychology and rehabilitation. In Licht, S. (ed.), *Rehabilitation and Medicine 1968*. New Haven: Elizabeth Licht, 1968, pp. 129–151.
3. Terry, R.D. Dementia: brief and selective review. *Arch. Neurol.* 33:1–3, 1976.
4. Miller, W.F., Taylor Pierce, A.K. Rehabilitation of disabled patient with chronic bronchial and pulmonary emphysema. *Am. J. Public Health* 53(Suppl.):18–24, 1963.
5. Gordon, E.E. Energy costs of activities in health and disease. *Arch. Intern. Med.* 101:702–713, 1968.
6. Lehneis, H.R. New developments in lower limb orthotics through bioengineering. *Arch. Phys. Med. Rehabil.* 53:303–310, 1972.
7. Hansson, J. Leg amputee: clinical follow-up study. *Acta Orthop. Scand.* 69:1–116, 1964.
8. Mashiocchi, C. Computerized Assessment Model for Rehabilitation Patients: A Program Evaluation of the Piersol Rehabilitation Center. RT27 Monograph 1982, National Institute of Handicapped Research, Department of Health, Education and Welfare.

The Aging Population: Prosthetic and Orthotic Considerations

▶

John H. Bowker, M.D.

19

As in all other areas of health care provision, the elderly person requires special consideration of his or her prosthetic and orthotic needs. Relative fragility of skin requires that artificial limbs and braces provide a safe and tolerable interface with the patient. The low forces producing fracture of major bones dictate that these devices be designed with inherent stability. Two types of disability that are very common in older persons will be discussed here in some detail to illustrate an approach to prosthetic and orthotic problems. These are lower limb amputation and the poststroke syndrome.

LOWER LIMB PROSTHESES

Virtually all amputations in the elderly are performed for dysvascular disease of the lower limbs. Trauma may contribute to loss of limb viability, but this usually occurs in the form of unnoticed microtrauma, especially in diabetic patients with partially or completely hypesthetic feet.

The heavy concentration of lower limb amputations in the elderly, added to their steadily increasing proportion of the population, has led to organizational and fiscal problems in provision of prosthetic services. Frustration with these obstacles has sometimes resulted in inadvertent exclusion of the elderly from prosthetic services on the basis of chronologic age, associated disease, high level of amputation, physical deconditioning, cost of services, or various combinations of these. Each of these factors should be considered not as reasons for exclusion, but as obstacles to be overcome, if at all possible.

Chronologic age should not be confused with physiologic age when considering fitting for prostheses. Many persons in their eighth and ninth decades are community ambulators with a single below-knee prosthesis, while most with bilateral below-knee prostheses can be household ambulators. More important than specific age is the person's general health and motivation toward independence.

Associated disease processes are to be expected in this age group. Many elderly amputees with uncompensated heart disease or diabetes mellitus have been successfully fitted and trained to walk following careful digitalization, diuresis, and control of diabetes. Nor should above-knee amputation necessarily preclude successful prosthetic fitting in the older person with selective use of modern prosthetic designs. Physical deconditioning resulting from usually unnecessary forced recumbency following amputation should be reversed by a careful program of graduated exercise and crutch or walker ambulation. This activity should gradually restore the patient's preamputation level of endurance in preparation for limb fitting.

The costs of prostheses and the attendant rehabilitative services are real factors in determining who should be fitted and trained to walk. It should be noted, however, that prosthetic rehabilitation, like any other, includes various levels of achievement. For example, some older patients return to virtually all their preamputation activities, while others may use their prostheses only to assist in the transfer from bed to chair and to toilet. Even though the second group sets and achieves a lesser goal than the first group, they nonetheless may be able to remain in their own homes with minimal assistance, avoiding the tremendous cost and inherent dependency of nursing home care.

From the preceding discussion it can be concluded that the easy road to management of the elderly person with lower limb amputation is that of exclusion from services on the basis of the factors considered. The more challenging road is that of providing the aging patient with the full range of state-of-the-art amputation and prosthetic services.

Preamputation

Ideal prosthetic management can be divided into three phases: preamputation, amputation, and postamputation. To group these three phases under prosthetic management may seem strange until it is recognized that the functional result—that is, the choice of community ambulation, household ambulation, or use of a wheelchair—is most often determined by the surgeon. The most important aspect of the preamputation phase, therefore, is the selection of the lowest possible amputation level commensurate with the disease process. On a subjective level, it has to do with a surgical attitude—a willingness to look for ways to preserve function by operating at more distal levels, even at the risk of having a few patients whose wounds do not heal primarily. On an objective level, this means the determination of level by any of several types of blood flow or tissue oxygenation studies now available. It also means consultation with the vascular surgeon in unclear cases to determine whether vessel reconstruction will either produce healing of an ulcer or allow amputation lower than that predicted by blood flow studies. By utilizing this approach, the surgical leap from gangrenous toe to crippling above-knee amputation can usually be avoided.

If at all possible, the patient should be introduced preoperatively to selected members of the prosthetic team, especially a peer counselor—that is, a person with a similar amputation who has made a successful adjustment. Preoperative training in the use of ambulatory aids can be helpful both in physical conditioning and in accustoming the patient to what he will be using in the immediate postoperative period.

Amputation
The success of the operative phase as related to healing potential is determined by the level selected, the design of the amputation, the care exercised in tissue handling, and the immediate postoperative wound management. The functional quality of the amputation, as reflected later in the patient's degree of activity, is directly related to the level selected. For instance, a transmetatarsal amputation really does not require a prosthesis, merely a shoe modified by a steel sole plate and a toe filler. An ankle disarticulation with preservation of the heel pad (Syme amputation) is superior to a below-knee level. It provides both an end-bearing residual limb with excellent proprioceptive function and leverage with little increase in energy consumption. A knee disarticulation is superior to an above-knee amputation for the same reasons. In amputations through the tibia or femur, a longer amputation provides better leverage and rotational control of the prosthesis.

The design of the amputation is of great importance. For example, skin flap survival and, therefore, primary healing in below-knee amputations are enhanced by use of a posterior myocutaneous flap. This is true because the skin of this multilayered flap receives a generous blood supply from its underlying muscle, whereas the anterior skin, lying directly on periosteum, does not. The supporting bone should be smoothly beveled anteriorly so that propulsive forces applied to the bony lever by stump muscles during prosthetic walking will not traumatize the overlying skin. Stabilization of stump tissues is accomplished by suturing opposing muscles to each other (myoplasty) or to the bone (myodesis). This serves to place the muscles at resting length for better strength and to center the bone in its muscular envelope. In above-knee amputations, especially, this technique will prevent subcutaneous drift of the femur, which is often followed by skin ulceration.

Because amputation in the older person is done largely for dysvascular disease, tissue handling technique during amputation is critical. Many otherwise technically good amputations have failed to heal because the skin has been closed under tension or because the surgeon has handled it with forceps; both of these errors devitalize the skin edges. This frequently leads to a higher revision with resultant decrease in prosthetic function. A well-padded, rigid plaster of Paris dressing applied to a below-knee amputation immediately postoperatively can have several beneficial effects. Perhaps the most important is prevention of a knee flexion contracture. These are very difficult to reverse and may prevent prosthetic fitting. This well-padded cast also controls postoperative edema and protects the wound from bedclothes and other minor trauma.

Postamputation

The postamputation phase of prosthetic management starts with selection of therapy, during which it should be determined whether the patient is really a candidate for a prosthesis. Therapy can often begin a day or two after surgery with walking between parallel bars. By the time the wound is soundly healed, at 2 to 4 weeks, the patient should have regained good balance and endurance. Patients who cannot progress to a walker or crutches during this period are rarely candidates for a prosthesis.

Following sound wound healing, a temporary or training prosthesis is applied. With increased weight bearing during gait training, the stump will decrease in volume, necessitating socket changes every 2 to 4 weeks. This method also provides a relatively inexpensive way to evaluate borderline prosthesis candidates. Temporary limbs can be applied at any amputation level but have their greatest value at the below-knee level. As soon as the stump stabilizes volumetrically, a permanent limb is prescribed.

Prescription of a definitive limb for the elderly patient

with lower limb amputation requires special consideration of two factors: weight and safety. Excess prosthesis weight may increase energy demands to the point where the patient's cardiopulmonary reserve may be insufficient for ambulation with the prosthesis. This weight can be kept to a minimum by use of new laminates that allow a thinner, lighter socket without sacrifice of strength. Ultralight sockets can also be molded from thermoplastics, such as polypropylene. Metal components made from new light alloys can also be prescribed. The latest prosthesis designs that are based on an endoskeletal system can result in considerable weight saving as well.

The question of safety is largely a matter of preventing the above-knee amputee from falling. Many elderly patients have difficulty in coordinating hip extension with heel strike during prosthetic gait, resulting in buckling of the conventional single-axis, friction knee joint. A safety knee that supports the patient's weight in a few degrees of flexion without buckling should be prescribed. Another important way of preventing falls is the provision of good gait training by an experienced physical therapist before the patient is allowed to take the prosthesis home.

Another important but often neglected aspect of care for the unilateral lower limb amputee is preservation of the remaining foot. The diabetic amputee is especially at risk because of decreased sensation from neuropathy. Extradepth shoes with heat-moldable polyethylene foam insoles, appropriate metatarsal pads, and deerskin uppers to accommodate toe deformities provide excellent protection in most cases.

UPPER LIMB PROSTHETICS

Very few upper limb amputations are done in the elderly. If an amputation has been carried out at the above-elbow level or higher, the patient is probably not a candidate for a prosthesis because of the complexity of the conventional prosthesis for these levels. On the other hand, a patient who has undergone below-elbow or wrist disarticulation and who has good physical and cognitive function should definitely be evaluated by use of a training limb. If successful, this patient may be fitted with a limb of either body-powered or myoelectric design. Elderly persons who have been good prosthesis wearers for many years may be considered for replacement of conventional limbs with myoelectric limbs to achieve greater comfort and to lessen the energy demand.

LOWER LIMB ORTHOSES

The most common indication for a lower limb orthosis (brace) is in the poststroke syndrome. Even though the individual has been determined to be a "safe" walker in that he or she has generally good spatial perception, proprioception, and strength and no strong synergy patterns, an orthosis may still be required for walking. With lack of knee control on the affected side, a "floor reaction" ankle-foot orthosis (AFO), with the foot set in slight plantarflexion, may be used. This forces the knee to extend fully in stance phase without bracing above the knee. If control of foot placement alone is lacking during stance phase and the patient has only mild equinovarus spasticity, a molded plastic AFO, trimmed to allow some dorsiflexion, is indicated. With moderate equinovarus spasticity, the AFO should have more anterior trim lines to capture the heel and thus control the subtalar joint. These plastic braces, molded individually for each patient, weigh only a few ounces.

If strong equinovarus spasticity persists throughout swing and stance phases, it is unlikely that positional control can be achieved by any sort of orthosis because the foot will invert and plantarflex inside the device. In this case, bracing may be preceded by transfer of the lateral half of the anterior tibial musculotendinous unit to the lateral side of the foot. Percutaneous lengthening of the Achilles tendon and of spastic toe flexors should be done at the same time, if indicated. Thereafter, foot control with a plastic AFO allowing dorsiflexion should be possible.

Braces that extend above the knee have no place in the treatment of stroke patients. Such devices are heavy and are virtually impossible to put on with one hand. If a stroke patient needs a locked knee to stand, he or she has insufficient limb control to walk safely.

UPPER LIMB ORTHOSES

In the elderly, the poststroke syndrome is again the most common condition requiring an orthosis. Simple devices, usually fabricated by the occupational therapist, are used to compensate for specific neuromuscular deficits or to control their effects.

Severe spasticity of the finger flexors can be controlled by a simple firm cone placed in the palm and secured to the hand with a dorsal Velcro strap. This device places the fingers in a resting position, which enhances flexor relaxation. The pan splint, frequently used in the past, requires that the fingers be forced into full extension and strapped to its flat surface. This induces increased flexor tone, causing the fingers to flex at the interphalangeal joints and extend at the metacarpophalangeal joints despite the straps. Patients find them intolerable.

Persistent inferior shoulder subluxation is often seen in the poststroke patient. It is due to flaccid paralysis of the deltoid and rotator cuff muscles and frequently becomes painful. A dynamic sling incorporating an elastic strap that approximates the humeral head to the glenoid may be prescribed. Use of a sling should be accompanied by an exercise program designed to prevent abduction–internal rotation contracture of the shoulder. At times, painful subluxation may be relieved even more simply by placing a rolled towel in the axilla and securing it with a figure-of-eight bandage, as is used for immobilization of clavicular fractures. If the patient with shoulder subluxation is wheelchair-bound, a trough made of wood or plastic can be attached to the armrest to support the forearm and reduce the shoulder.

CONCLUSIONS

The prosthetic and orthotic needs of the elderly are related largely to the high incidence of generalized vascular disease in this population, which results in the amputation of dysvascular lower limbs and in the poststroke syndrome. There are many obstacles to the provision of prosthetic services to the elderly, some real, including physical or mental inability to use a prosthesis, and some perceived, such as advanced chronologic age and associated disease processes.

Ideal prosthetic management includes attention to the preamputation and amputation phases as well as to the more obvious postamputation phase. This is necessary because the surgeon selects the amputation level, which is the major determinant of the amputee's future level of functioning. A method of determining candidacy for prosthetic fitting and training will help avert costly errors in the prescription of prostheses for patients who are unable to use them.

Elderly persons with hemiplegia following a cerebrovascular accident may be able to walk with an ankle-foot orthosis if they meet certain criteria that indicate they will be "safe" walkers. Braces that extend above the knee have no place in this disorder. If the upper limb exhibits severe spasticity and/or shoulder subluxation, the occupational therapist can construct and apply any of several simple orthotic devices.

This chapter emphasizes the uniqueness of the elderly population group in the application of basic prosthetic and orthotic principles. Readers wishing further information should refer to the bibliography.

BIBLIOGRAPHY

American Academy of Orthopaedic Surgeons. *Atlas of Limb Prosthetics: Surgical and Prosthetic Principles.* St. Louis: C.V. Mosby, 1981.

American Academy of Orthopaedic Surgeons. *Atlas of Orthotics: Biomechanical Principles and Application.* St. Louis: C.V. Mosby, 1975.

Bowker, J.H. (ed.). Symposium on special problems in orthopedic rehabilitation. *Orthop. Clin. North Am.* 9(2): April 1978.

Farber, S.D. *Neurorehabilitation: A Multisensory Approach.* Philadelphia: W.B. Saunders, 1982.

Waters, R.L. (ed.). Symposium: Stroke rehabilitation. *Clin. Orthop.* 131: March-April 1978.

Index

A

Accidents, *12–15*
  deaths from, *12, 141*
  factors in, *13–15*
  incidence of, *141*
  motor vehicle, *14, 143*
  surgical approaches to injuries, *144*
Acid-base challenge, and exercise, *22*
Acromegaly, *87*
Aerobic training
  age and, *94–95*
  cross-sectional studies of, *92–94*
Aging
  biological nature of, *24*
  characteristics of, *17–23*
  definition of biological, *9*
  demographics of, *1–8, 9–10*
  disease and, *22–23*
  experimental approaches to study of, *24*
  food restriction and, *24–30*
  mortality and, *17–18*
  pathoanatomy of, *33–35*
  physiologic deterioration and, *21–22*
  psychosocial factors in, *9–15*
Alcohol, and accidents, *13*
Aldohistidine, and collagen aging, *42*
Amputation, lower limb, *165–166*
Androgens, and osteoporosis, *87*
Ankle-foot orthosis (AFO), *168, 169*

Arachidonic acid, and cartilage aging, *61*
Articular cartilage, *59–63*
  aging and, *115–116*
  bone in joint function and, *128*
  calcification of, *125*
  cell counts in, *59–60*
  collagen in, *60*
  mechanisms by which nutrients enter, *122*
  nutritional changes at base of, *121–126*
  osteoarthritis and, *114, 115–116*
  proteoglycans and, *60–61*
  repair of, *116*
  stiffness gradients and damage to, *130*
  water content in, *60*
Atherosclerosis
  aging-independent, *34, 35*
  cardiac functional changes with, *91–92*
  pathogenesis of, *113*
Automobile accidents, *14, 143*

B

Basement membrane
  collagen of, *38*
  muscle aging and, *53*
Bile duct hyperplasia, *28*
Blacks, longevity among, *12*. *See also* Race
Body composition
  aging and, *18–21*
  food restriction and aging and, *26–27*
Body weight, and aging, *19–20*

171

Index

Bone
  aging and, 45, 65–74
  areas of loss in, 71–72
  assessment methods for, 65–66
  calcium-to-phosphorus ratio in, 84
  collagen aging and, 38, 44–45
  cortical dimension changes in, 67–68
  exercise and, 81
  fluoride and mass of, 89
  fractures in osteoporosis and, 76
  intracortical porosity changes in, 68–69
  joint function and, 128–130
  lumbar spinal, 107–108
  malnutrition and demineralization of, 15
  mechanism of loss of, 66–69
  mineral and collagen in, 45
  morbidity and amount of, 45
  osteoporosis and mass of, 80–81
  phosphorus and loss of, 83–84
  plugging and calcification of, 70–71
  race and sex and aging of, 66
  remodeling of, 69–73
  treatment of loss of, 84–85
Bone marrow, and joint nutrition, 122
Bureau of the Census, 2
Burns, 142

C

Calcitonin, and osteoporosis, 86
Calcium
  bone mass and, 81
  calcitonin and, 86
  control of, 81–82
  deficiency of, 83
  muscle aging and, 53
  osteomalacia and, 35
  osteoporosis and, 34, 78, 80, 81–83, 86, 88
  phosphorus ratio to, 84
  sources of, 82–83
Calcium pyrophosphate, and osteoarthritis, 117
Cardiac fibrosis, 28
Cardiac functional changes, 91–94
Cartilage, articular. *See also* Articular cartilage
  osteoarthritis and, 34
Central nervous system
  lumbar spine and, 108
  trauma and aging and, 49–50
Cervical spine fractures, 144
Cholesterol, age-related increases in, 27
Chondrocalcinosis, 34
Chondroitin-4-sulfate, and osteoarthritis, 34
Chondroitin sulfate A and C, and articular cartilage, 60
Collagen, 37–46
  aging of, 37–46
  articular cartilage aging and, 60
  biosynthesis of, 43
  bone and, 44–45
  dissociative solvents in study of, 40–41
  electron microscopy of, 38–39
  freeze-fracture-etching of, 39–40
  intervertebral disk and, 104–105
  light microscopy of, 38
  mechanical properties of, 44–45
  metabolic turnover in, 43–44
  mineral in bone and, 45
  morphologic observations of, 38–41
  nonreducible cross-links in, 42–43

Index

plastic strain of cortical bone and, *108*
reasons for aging of, *37–38*
reducible cross-links in, *41–42*
response to trauma and repair by, *45–46*
structure and chemistry of, *41–43*
types of, *38*
Colles' fracture, *75, 76, 132*
Color perception, and accidents, *13–14*
Conduction velocity, and aging, *95*
Conference on the Epidemiology of Aging, Second, *9–10*
Corticosteroids, and osteoporosis, *34*

D

Death rates. *See* Longevity; Mortality
Demineralization of bone, *15*
Demographics of aging, *1–8, 9–10*
  aging of population, *2*
  economic constrains and issues, *3–5*
  older persons continuing in active labor force, *3*
  older persons in proportion to all other younger adults, *2*
  political factors and, *5–8*
  proportion of older persons in population, *1–2*
Depression, mental, *13, 157*
Depth perception, and accidents, *14*
Diet
  aging and food restriction, *24–30*
  bone mass and, *81*
  calcium and, *81, 82–83*
  hip replacement and, *138–139*
  longevity and, *11*

Diffuse idiopathic skeletal hyperostosis (DISH), *41–42*
Dihydroxylysinonorleucine, and collagen aging, *41–42*
1,25-Dihydroxyvitamin D
  calcium absorption and, *81*
  osteoporosis and, *34, 87–88*
Diphosphonates, and osteoporosis, *86*
Disease
  accidents and, *13*
  aging and, *22–23*
  aging-concomitant, *33–34*
  aging-dependent, *34*
  aging-independent, *34–35*
  food restriction and delay in, *28*
Disk, intervertebral, *103–106*
  aging changes in, *104*
  biochemical changes with age in, *105, 106*
  collagen fibers in, *104–105*
  degeneration of, *109–110*
  herniation of, *110*
Drugs, and accidents, *13*
Duke (University) Longitudinal Study, *11*

E

Economic factors
  demographics of aging and, *3–5*
  federal programs and, *4–5*
  natural resources availability and, *3*
  productivity and expansion and, *4*
Electron transport, and food restriction, *29–30*
Estrogen, and osteoporosis, *78, 80, 87*
Exercise
  aerobic training and, *92–95*
  aging and, *21*
  bone loss and, *81*
  longevity and, *11*

173

Exercise *(continued)*
  muscular strength changes with aging and, 96–99
  strength training and, 99–100
Eyesight, and accidents, 13–14

F

Facet joint of spine, 106–107
Falls, 14, 141–142
Family, and rehabilitation program, 159, 160
Fat mass
  aging and, 20
  food restriction and aging and, 26–27
Federal Aviation Agency, 54–55
Federal programs for the elderly, 4–8
Femur fractures, 75, 76, 147
  morbidity with, 149–150
  Singh index for, 77
Fibroblast growth factor, 118
Fibrosis, cardiac, 28
Fires, 14, 140
Fluoride, and osteoporosis, 88–89
Food. *See* Diet
Fractures, 144
  hip replacement in, 135
  osteoarthritis and, 132–133
  osteoporosis and, 75–76
  treatment of, 84–85, 89
Framingham Study, 20

G

Glucagon, and age-related decline, 27
Growth hormone, and osteoporosis, 86–87

H

Haversian canals, 70, 72, 73
Head injuries, 144
Health care providers, 8
Hearing loss, and accidents, 13

Heberden's nodes, with osteoarthritis of hip, 116
Heredity, and longevity, 12
Hip, Heberden's nodes with osteoarthritis of, 116
Hip fractures, 147–154
  classification of patients with, 151
  femoral neck, 149–150
  indicators of recovery following surgery for, 151
  intertrochanteric fractures in, 150
  morbidity with, 149–150
  mortality with, 147–149
  rehabilitation in, 151–153
  treatment recommendations for, 151–153
Hip replacement, 135–140
  joint sepsis and, 136–137
  loosening in, 137–138
  numbers of, 135
  nutritional status and, 138–139
  patient population in, 135–136
  porous-coated implant in, 138
  problems related to, 136–139
  prospectus for, 139
  reasons for, 135
  results of, 136
  specially designed prostheses in, 138
  thromboembolic disease in, 137
  tumors and, 138
Histidinoalanine, and collagen aging, 42–43
Histidinohydroxymerodesmosine, and collagen aging, 42
Hormones
  food restriction and aging and, 28
  osteoporosis and, 87
Human services professionals, 8
Hyaluronic acid, and articular cartilage aging, 60

Hydroxymerodesmosine, and collagen aging, 42
Hydroxyapatite, and osteoarthritis, 117
Hypercortisonism, 77
Hyperparathyroidism, and osteoporosis, 77, 88
Hyperthyroidism, 77
Hypothalamus, and aging, 24

I
Incontinence, 160
Intelligence, and longevity, 12
Interstitial collagen, 38

J
Joint function, and bone, 128–130
Joint sepsis, in hip replacement, 136–137

K
Keratosulfate, and articular cartilage, aging, 60, 61
Kidney function, 22
Kinematics of aging, 91–100
  aerobic training with age and, 94–95
  cardiac functional changes and, 91–94
  muscular strength changes in, 96–99
  musculoskeletal and neuromuscular changes in, 95–96
  strength training and, 99–100

L
Labor force, older persons in, 3
Lipid metabolism, 28
Longevity, 11–12
  body mass and, 20–21
  factors affecting, 11–12
  food restriction and, 24–30
  predictors of, 12
  sex differences in, 10, 12

Lower limbs
  orthoses for, 168
  prostheses for, 163–167
Lumbar spine aging, 103–111
  disk degeneration in, 109–110
  disk herniation in, 110
  epidemiology of, 103
  osteoporosis of, 109
  prevention and treatment of age-related changes in, 111
  stenosis of, 110–111
Lysinoalanine, and collagen aging, 42
Lysozyme, and articular cartilage aging, 60
Lysol oxidase, and collagen aging, 43

M
Malnutrition, and smell decrease, 14–15
Marital status, and longevity, 11
Medicaid, 6
Medicare, 6
Men. *See* Sex differences
Morbidity
  amount of bone and, 65
  trauma and, 143–144, 145
Mortality. *See also* Longevity
  accidents and, 12–15, 141
  aging and, 17–18
  early retirement and, 10
  hip fractures and, 147–149
Motor vehicle accidents, 14, 143
Muscle
  anatomic changes in, with aging, 52–53
  food restriction and aging and, 28
  histochemical studies of aging of, 53–54
  kinematics of aging and, 95–96
  lumbar spinal, 108
  performance capability and, 54–56

Muscle *(continued)*
  quantitative aspects of aging of, *49–54*
  strength and aging in, *96–99*
  strength training and, *99–100*
Myofibril atrophy, *33*
Myosin adenosine triphosphatase (ATPase), and muscle aging, *54*

N

National Institutes of Health, *135*
Nephropathy, chronic, *28*
Neural tissues
  strength training and, *100*
  trauma and aging of, *49–50*
Neuromuscular system, and aging, *95–96*
Nursing homes, *141–142*
Nutrition. *See* Diet

O

Older persons
  aging of aging population and, *2*
  continuing in active labor force, *3*
  in proportion to younger adults, *2*
  proportion of, in population, *1–2*
Orthoses
  lower limb, *168*
  upper limb, *168–169*
Osteoarthritis
  articular cartilage repair in, *116*
  biomechanics of, *127–133*
  bone in joint function and, *128–130*
  bone lips and spurs on peripheral margins in, *132–133*
  cartilage aging and, *61*
  clinical complaints in, *117–118*
  development of, *131–133*
  energy absorption and joint congruence in, *130, 131*
  etiology of, *115–116*
  fractures and, *132–133*
  hip replacement in, *135*
  osteoporosis and, *133*
  pathogenesis of, *113–119*
  pharmacologic management of, *118*
  primary, *34*
  as remodeling process, *114–115*
  secondary, *34–35*
  stiffness gradients and cartilage damage in, *130*
  synovial fluid crystals in, *117*
  terminology used in, *127*
  variant forms of, *116–117*
Osteoarthrosis, *127*. *See also* Osteoarthritis
Osteomalacia, *34–35*
Osteopetrosis, *68*
Osteoporosis, *75–90*
  agents that worsen, *85–87*
  aging-concomitant, *33, 34*
  bone loss mechanisms in, *68*
  bone mass and, *80–81*
  calcium and, *81–83*
  calcium-to-phosphorus ratio in, *84*
  causes of, *78, 89–94*
  definition of, *75–76*
  idiopathic, *78*
  osteoarthritis and, *133*
  pathology and physiology of, *78–79*
  prevention of, *87–88*
  radiologic appearance of, *76–77*
  Singh index in, *76–77*
  spinal, *109*
  treatment of, *84–89*
  types of, *77–78*
Oxygen uptake, maximal ($\dot{V}O_2$max)
  aging and, *21*
  cardiac functional changes and, *92–94*

Index

P
Paget's disease, 86
Parathyroid hormone
  calcium and, 81–82, 84
  osteoporosis and, 34
Pelvis fractures, 144
Performance capability, and muscle aging, 54–56
Phosphates, and osteoporosis, 85–86
Phosphorus
  bone loss and, 83–84
  calcium ratio to, 84
  osteomalacia and, 35
  osteoporosis and, 78, 80, 83–84
  parathyroid hormone and, 82
Physical therapy, 144, 160
Physiologic deterioration with aging, 21–22
Political factors, and demographics of aging, 5–8
Population. See Demographics of aging
Poststroke syndrome, 168, 169
Prostheses
  amputation and, 165–166
  hip replacement surgery with, 138
  lower limb, 163–167
  postamputation period and, 166–167
  preamputation period and, 164–165
  upper limb, 167
Proteoglycans, and articular cartilage aging, 60–61
Psychosocial factors in aging, 9–15
  accidents among older persons and, 12–15
  involuntary relocation and, 10
  longevity and, 10, 11–12
  mortality after early retirement and, 10
  suicide and, 10
Pyridinoline, and collagen aging, 42

R
Race
  bone aging and, 66
  bone mass and, 80
  longevity and, 12
Reagan administration, 6
Rehabilitation medicine, 157–160
  goals of, 159–160
  hip fractures and, 150–151
  program in, 157–159
  prosthetic and orthotic considerations in, 163–170
  surgery after accidents and, 144
Relocation, involuntary, 10
Retirement, and longevity, 10, 11
Rheumatoid arthritis, 135
Rib fractures, 144

S
Sarcoplasmic reticulum, and aging of muscle, 53
Sex differences
  bone aging and, 66
  bone mass and, 80
  longevity and, 10, 12
Singh index, 76–77
Skeletal muscle
  anatomic changes with aging in, 52–53
  food restriction and aging and, 28
  histochemical studies of aging in, 53–54
Skin, and collagen aging, 38, 40
Smell decreases, and accidents, 14–15
Smoking, and longevity, 11
Smooth muscle, 28
Social activity, and longevity, 11–12
Social Security programs, 4, 5, 6, 8
Social services programs, 6–7
Socioeconomic status, and longevity, 12

Spine. *See also* Lumbar spine
 age-specific physical changes in, *103, 104*
 bone of, *107–108*
 components of, *103–108*
 disks of, *103–106*
 facet joint of, *106–107*
Spondylosis deformans, *115*
Sports
 central nervous system aging and, *50*
 muscle aging and performance in, *55–56*
Strength changes, and aging, *96–99*
Strength training, *99–100*
Stroke, *168, 169*
Subchondral bone, and joint function, *128–129*
Suicide rates, *10, 136*
Superoxide dismutase, and food restriction, *29–30*
Survival curves, and aging, *18*
Syme amputation, *165*
Synovial fluid, and joint nutrition, *122*
Synovitis, *117*

T

Tactile sensation loss, and accidents, *13*
Thromboembolic disease, and hip replacement, *137*
Traffic accidents, *14, 143*
Trauma, *141–145*
 articular cartilage and, *61–62*
 burns and fires and, *142*
 central nervous system aging and, *49–50*
 collagen aging and, *45–46*
 definition of, *141*
 falls and, *141–142*
 motor vehicle accidents and, *14, 143*
 stiffness gradients and cartilage damage in, *130*
 trabecular bone in energy absorption in, *108*
Tumors
 food restriction and, *28*
 joint replacement surgery and, *138*
Turner's syndrome, *78*

U

Upper limbs
 orthoses for, *168–169*
 prostheses for, *167*

V

Vertebral fracture, *75*
Vitamin D. *See also* 1,25-Dihydroxyvitamin D
 osteomalacia and, *35*

W

Weight, body, and aging, *19–20*
Whites. *See also* Race
 longevity among, *12*
Women. *See also* Sex differences
 calcium deficiency in, *83*
 muscular strength changes with aging in, *96–99*
 osteoporosis in, *78*